CESMM4

Civil Engineering Standard Method of Measurement

Fourth edition

Published by ICE Publishing, 40 Marsh Wall, London E14 9TP

Full details of ICE Publishing sales representatives and distributors can be found at:
www.icevirtuallibrary.com/printbooksales

www.icevirtuallibrary.com

A catalogue record for this book is available from the British Library

ISBN 978-0-7277-5751-7

ICE Publishing is a division of Thomas Telford Ltd, a wholly-owned subsidiary of the Institution of Civil Engineers (ICE).

Senior Commissioning Editor: Gavin Jamieson
Production Editor: Imran Mirza
Market Development Executive: Catherine de Gatacre

Typeset by Academic + Technical, Bristol
Printed and bound in Great Britain by CPI Group (UK) Ltd, Croydon, CR0 4YY

Civil Engineering Standard Method of Measurement

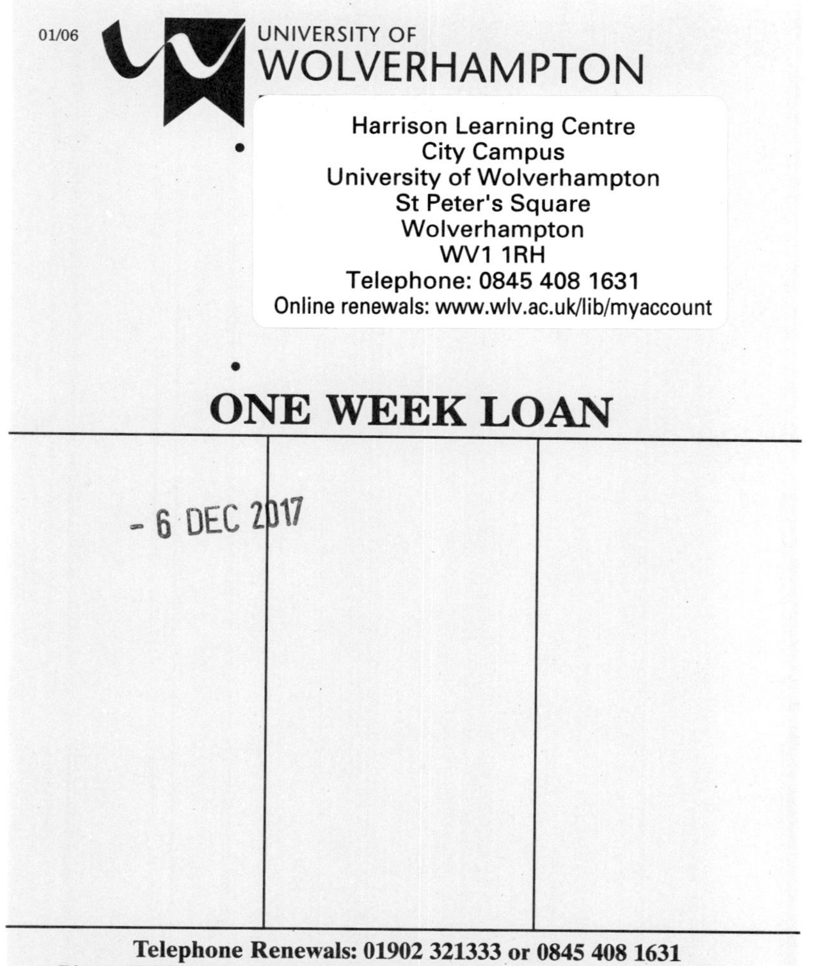
01/06
UNIVERSITY OF WOLVERHAMPTON
Harrison Learning Centre
City Campus
University of Wolverhampton
St Peter's Square
Wolverhampton
WV1 1RH
Telephone: 0845 408 1631
Online renewals: www.wlv.ac.uk/lib/myaccount
ONE WEEK LOAN
- 6 DEC 2017
Telephone Renewals: 01902 321333 or 0845 408 1631
Please RETURN this item on or before the last date shown above.
Fines will be charged if items are returned late.
See tariff of fines displayed at the Counter. (L2)

Contents

Preface

The *Civil Engineering Standard Method of Measurement*, fourth edition (CESMM4), has been approved by the sponsors – the Institution of Civil Engineers and the Civil Engineering Contractors Association – for use in works of civil engineering construction. This fourth edition supersedes the third edition published in 1991.

CESMM4 may be used with any conditions of contract for civil engineering work that includes measurement.

The Committee under whose guidance CESMM4 has been prepared will keep the use of the document under review and consider any suggestions for amendment. These should be addressed to The Secretary, The Institution of Civil Engineers, 1–7 Great George Street, London SW1P 3AA. Revision of the document will be made when such action seems warranted.

Foreword to the first edition

In 1964, the Council of the Institution of Civil Engineers set up a Committee under the Chairmanship of T. A. L. (now Sir Angus) Paton, CMG, BSc(Eng), FICE, to propose revisions to the *Standard Method of Measurement of Civil Engineering Quantities*. In 1971 the work of revision was undertaken by Martin Barnes, PhD, BSc(Eng), MICE, who worked initially under an agreement with the University of Manchester Institute of Science and Technology and, after he left the University in 1972, under a direct agreement with the Institution of Civil Engineers. A Steering Committee was appointed by the Council of the Institution of Civil Engineers to supervise the work. The Steering Committee was enlarged in 1972 by the appointment of an additional representative from each of the Institution of Civil Engineers, the Association of Consulting Engineers and the Federation of Civil Engineering Contractors. The members of the committee (*indicates member appointed in 1971) were

*D. C. Coode, FICE, FIEAust, Chairman
*M. Agar, BSc, FICE, FIStructE, Institution of Civil Engineers
F. J. Cave, BSc, FICE, FRICS, MTPI, FIMunE, FRSH, Institution of Civil Engineers
*H. R. Oakley, MSc(Eng), FICE, MIWE, FASCE, Association of Consulting Engineers
P. B. Ahm, MSc, FICE, Association of Consulting Engineers
*R. B. Hill, BSc, FICE, FIStructE, Federation of Civil Engineering Contractors
J. A. Sneden, FRICS, FIQS, Federation of Civil Engineering Contractors
J. B. B. Newton, BSc(Eng), FICE, co-opted
N. C. B. Brierley, BSc(Eng), FICE, co-opted

Late in 1972, a draft of the revised standard method of measurement was circulated for comment to 71 representative bodies. Trial Bills of Quantities were prepared by 20 organizations. Discussions with interested bodies continued throughout 1973 and 1974 and the form of the present document owes much to the suggestions made during this time by bodies outside the committee. The Steering Committee is indebted to the many people who helped in this way.

The object of the work has been to make improvements while retaining the good features of the previous edition of the standard method of measurement. The principal improvements sought are

(*a*) to standardize the layout and contents of Bills of Quantities prepared according to the standard method of measurement
(*b*) to provide a systematic structure of bill items leading to more uniform itemization and description
(*c*) to review the subdivision of work into items so that a more sensitive and balanced description of the value of work in a contract is provided
(*d*) to take account of new techniques in civil engineering construction and management, their influence on the work itself and on the administration of contracts.

A Bill of Quantities which in essence is no more than a price list of the Permanent Works no longer adequately reflects the many variables in the cost of civil engineering construction which have resulted from developments in constructional techniques and methods. It has therefore been decided to provide for some additional items of measured work and for other items, entered at the option of the tenderer, directly related to methods of construction.

A system of work classification has been adopted as the basis of the method of measurement so that Bills of Quantities can be compiled and used more easily. The system should enable much of the repetitive clerical work associated with the use of Bills of Quantities to be simplified, and make the use of computers easier.

The Work Classification incorporates a reference number for each type of work component. These reference numbers may be used as a simple code for identification of work. Their use as part of the item numbers in Bills of Quantities is suggested, but is optional. The coding is sufficiently flexible not to inhibit description of the particular work in each contract.

Foreword to the second edition

In 1983, the Council of the Institution of Civil Engineers instructed the CESMM Review Committee to prepare a second edition of the *Civil Engineering Standard Method of Measurement*. The members of the Review Committee were

H. R. Oakley, CBE, FEng, MSc(Eng), FICE, Chairman
D. C. Coode, CBE, FCGI, FICE
N. C. B. Brierley, BSc(Eng), FICE
N. M. L. Barnes, BSc(Eng), PhD, FICE, FCIOB, ACIArb, MBCS

The work of analysing the comments received on the first edition and of drafting the second edition was undertaken by Martin Barnes and Partners who were assisted by McGill and Partners.

Many organizations contributed comments and suggestions for amendment of the document and others participated by reviewing drafts and giving advice. The Review Committee is indebted to the many people who helped in this way, and particularly to the Federation of Civil Engineering Contractors who appointed a committee to review and comment upon the drafts.

The object of the amendments made in the second edition was twofold. Firstly it was to take account of developments in civil engineering technology and in the significance of different costs of civil engineering work which have taken place since the first edition was published. Secondly it was to take the opportunity of altering the wording of the small number of provisions of the first edition which experience had shown were not working either as smoothly as they might or in the way which had been intended. The second edition of the CESMM is consequently not a radical departure from the first edition, but an update and general overhaul.

The two most noticeable changes are the categorization and upgrading of the former notes in the work classification and the introduction of a standard method of measurement for sewer renovation work. The rearrangement of the former notes is intended to make the document easier to use in two ways. Firstly, the notes have been re-named rules in order to emphasize that their provisions govern how work should be described and measured in civil engineering bills of quantities and that they have equal status with rules in any other part of the document. Secondly, they have been divided into four categories to indicate the four separate functions which they perform and have, as far as possible, been laid out alongside the parts of the classification tables to which they relate.

The section for measurement of sewer renovation work has been introduced in order to provide for the considerably increased volume of work of this type now being undertaken. The method of measurement for sewer renovation in this document is based upon that devised originally for the Water Research Centre and included in their standard specification. The Institution of Civil Engineers and the Federation of Civil Engineering Contractors acknowledge the assistance given by the Water Research Centre in allowing this method of measurement to be embodied in the CESMM.

Foreword to the third edition

Although the second edition of the *Civil Engineering Standard Method of Measurement* was published in 1985, a number of factors have combined to cause the need for a third edition. The main changes incorporated in the third edition (CESMM3) are amendments to enable its use with the ICE Conditions of Contract, sixth edition (January 1991), and the introduction of a standard method of measurement for water main renovation and for simple building works incidental to civil engineering works (classes Y and Z). Account has also been taken of developments in civil engineering technology and the opportunity taken to make minor amendments and corrections to the text of the second edition. There are no changes in principle or in the general arrangement.

The Review Committee is indebted to a number of organizations and individuals who have contributed comments and suggestions for amendment. The committee is particularly indebted to E. C. Harris: Quantity Surveying whose proposals formed the basis for the new class Z. The work of analysing the comments and suggestions and drafting CESMM3 was undertaken by Coopers & Lybrand Deloitte.

The members of the Review Committee were

H. R. Oakley, CBE, FEng, MSc(Eng), FICE, Chairman
N. M. L. Barnes, FEng, BSc(Eng), PhD, FICE, FCIOB, ACIArb, MBCS, CBIM
Miss R. Beales, Barrister-at-Law
J. Banyard, MICE
H. A. Jones, MICE, Secretary

The committee is indebted to Dr Barnes of Coopers and Lybrand Deloitte for his invaluable advice and assistance throughout.

Foreword to the fourth edition

Since its introduction in 1976, the Civil Engineering Standard Method of Measurement (CESMM) has proved to be extremely resilient in meeting the needs of those engaged in preparing contracts based on traditional 'measure & value' principles. There were updates in 1985, and 1991, which were largely concerned with introducing new Work Classifications for renovation of water mains and sewers, together with updating to accord with revised versions of the ICE Conditions of Contract; a little tidying up to deal with specific problems that had emerged; and updating references to British Standards contained in the Work Classification.

However by 2009 it was being suggested that CESMM required updating to meet current engineering practice and procedure, and a review was initiated by the Institution of Civil Engineers. The review concluded that an update was indeed necessary.

The industry had changed substantially since CESMM had been introduced over 35 years ago. When it was first published, most work was carried out under the ICE Conditions of Contract, or FIDIC, which were broadly similar to one another. Today there is a variety of contracts for civil engineering including NEC and the Infrastructure Conditions of Contract (formerly known as the ICE Conditions of Contract). Additionally some Clients have adopted standard forms produced by other engineering institutions, and large projects are sometimes let on bespoke contracts drawn up by specialist lawyers, which may or may not have some similarity to one or other of the many 'standard forms' now available.

The challenge of updating references to technical standards had been identified, and is increased by the use of CESMM in other countries which have their own national standards and for whom British Standards are no longer relevant. Therefore, a decision was taken to remove most references to British Standards. As a consequence if Standards are used these have to be clearly identified in the contract documents.

Finally the review identified that in some cases, although the main contract may well be lump sum or target cost, contractors use CESMM for tendering purposes, and in some cases let sub-contracts on the basis of re-measurement, not withstanding that the main contract is not on a remeasurement basis.

To deal with the issues raised by these practices, since there is far less standardisation than there was when CESMM was first published, two major changes to this 4th Edition of CESMM have been introduced:

1. The document is 'contract neutral'. That is to say it does not depend on any particular form of contract, but of course the contract must accommodate a measure & value approach. It is therefore necessary for the Bill compiler to identify in the Preamble the relevant clauses within the conditions of contract adopted. Additionally the terminology and responsibility of individuals varies from contract to contract (e.g. terms such as 'The Engineer to the Contract' have been superseded in some standard forms), but the responsibilities and designation of these individuals within the Method of Measurement must accord with the chosen contract. Therefore a schedule has been introduced into the Preamble as a mandatory requirement to ensure that this need for consistency between the contract documents and the Method of Measurement is not overlooked.

2. The document is generally 'National Standard neutral'; that is, there is limited reference to British or other national standards and that information must be given elsewhere on the drawings or in the specification. There are two exceptions to this principle where it has proved impossible to produce Bill items without reference to some form of standard classification. These are:

   a. Concrete mixes
   b. Road construction.

   In these two cases the latest British Standards and Specification for Highways Works available from The Stationery Office have been adopted, including the dates of publication. It is essential that when billing this type of work compilers do check that these standards are indeed the ones to which they wish to refer to and, if not, complete the schedule in the Preamble and modify the item descriptions accordingly. It is simply not possible to republish CESMM every time that a standard is updated.

The opportunity has also been taken to update a number of other areas, and deal with a few known problems that have emerged over the years.

A CESMM advisory service has been established since publication of CESMM and will continue for CESMM4. However, it is not in a position to advise on derivatives of CESMM.

Finally CESMM4 retains essentially the same principles as when the document was first published, and it is a great tribute to the authors of that first edition that they were able to produce a methodology that has proved so resilient to the changing practices and procedures within the construction industry. It is the intention of the Review Committee that this edition will serve its users as well as the earlier editions have done.

The Review Committee received advice and helpful comments from a wide range of organisations and individual practitioners, and would like to express its gratitude to all of them for the assistance they provided.

The drafting of CESMM4 was undertaken by Richard McGill; for the update of Class S Rail Track assistance was provided by Network Rail, who attended the latter meetings of the Committee.

The members of the Review Committee were

J. K. Banyard OBE, FREng, FCGI, BSc(Eng), FICE, FCIWEM
R. Beales, Barrister at Law
W. Edwards BSc, Dip Arb, FICE, FCIHT, MCInstCES, FCIArb
T. Pemberton MA, MLit
J. Fiske MACostE, BSc(Hons)
R. E. N. McGill FRICS
R. Gerrard BSc(Hons), FRICS, FCIArb, FCInstCES
P. Schwanethal FCInstCES, MCIArb
Secretariat: J. Hawkins MSc
S. Hernandez BSc, MSc

Civil Engineering Standard Method of Measurement
ISBN 978-0-7277-5751-7

doi: 10.1680/cesmm.57517.001

Section 1
Definitions

1.1. In this document and in Bills of Quantities prepared according to the procedure set forth herein the following words and expressions have the meanings hereby assigned to them, except where the context otherwise requires.

1.2. The word 'work' includes work to be carried out, goods, materials and services to be supplied, and the liabilities, obligations and risks to be undertaken by the contractor under the contract.

1.3. The contract administrator may be the employer, his agent or representative.

1.4. The expression 'expressly required' means shown on the drawings, described in the Specification or instructed by the contract administrator pursuant to the contract.

1.5. 'Bill of Quantities' means a list of items giving brief identifying descriptions and estimated quantities of the work comprised in a contract.

1.6. 'Daywork' means the method of valuing work on the basis of time spent by the operatives, the materials used and the plant employed.

1.7. 'Work Classification' means the Work Classification set out in section 8.

1.8. 'Original Surface' means the surface of the ground before any work has been carried out.

1.9. 'Final Surface' means the surface indicated on the drawings to which excavation is to be carried out.

1.10. 'Commencing Surface' means, in relation to an item in a Bill of Quantities, the surface of the ground before any work covered by the item has been carried out. 'Commencing Surface' means, in relation to a group of items in a Bill of Quantities for work in different materials in an excavation or a bored, drilled or driven hole, the surface of the ground before any work covered by any item in the group has been carried out.

1.11. 'Excavated Surface' means, in relation to an item in a Bill of Quantities, the surface to which excavation included in the work covered by the item is to be carried out. 'Excavated Surface' means, in relation to a group of items in a Bill of Quantities for excavation in different materials, the surface to which excavation included in the work covered by any item in the group is to be carried out.

1.12. A hyphen between two dimensions means a range of dimensions which includes all dimensions exceeding that preceding the hyphen but not exceeding that following the hyphen.

Section 2
General principles

Title application and extent

2.1. The title of this document is the *Civil Engineering Standard Method of Measurement,* fourth edition, which is abbreviated to CESMM4. CESMM4 is intended to be used only for civil engineering works and simple building works incidental to civil engineering works.

2.2. CESMM4 provides for simple building works incidental to civil engineering works to be measured in accordance with Class Z. CESMM4 does not deal with the preparation of Bills of Quantities for complex mechanical or electrical engineering work, or complex building work or work which is seldom encountered in civil engineering contracts. Where any such work is to be included in a contract for civil engineering work, it shall be itemized and described in the Bill of Quantities in sufficient detail to enable tenderers to price it adequately and the method of measurement shall be stated in the Preamble to the Bill of Quantities in accordance with paragraph 5.4.

Object of CESMM4

2.3. The object of CESMM4 is to set forth the procedure according to which the Bill of Quantities shall be prepared and priced and the quantities of work expressed and measured.

Objects of the Bill of Quantities

2.4. The objects of the Bill of Quantities are

(*a*) to provide such information of the quantities of work as to enable tenders to be prepared efficiently and accurately
(*b*) when a contract has been entered into, to provide for use of the priced Bill of Quantities in the valuation of work executed.

2.5. In order to attain these objects, work should be itemized in the Bill of Quantities in sufficient detail for it to be possible to distinguish between the different classes of work, and between work of the same nature carried out in different locations or in any other circumstances which may give rise to different considerations of cost. Consistent with these requirements the layout and content of the Bill of Quantities should be as simple and brief as possible.

2.6. All work which is expressly required should be covered in the Bill of Quantities.

2.7. CESMM4 seeks to attain these objects principally by the use of the Work Classification. This defines

(*a*) how work is to be divided into separate items in the Bill of Quantities
(*b*) the information to be given in item descriptions
(*c*) the units in which the quantities against each item are to be expressed
(*d*) how the work is to be measured for the purpose of calculating quantities.

Section 3
Application of the work classification

Item descriptions

3.1. The Work Classification divides work commonly encountered in civil engineering contracts into 26 main classes. Each class comprises up to three divisions which classify work at successive levels of detail. Each division comprises a list of up to eight descriptive features of work. Each item description in the Bill of Quantities shall identify the component of work covered with respect to one feature from each division of the relevant class, for example

> Class H (precast concrete) contains three divisions of classification. The first classifies different types of precast concrete units, the second classifies different units by their dimensions, and the third classifies them by their mass. Each item description for precast concrete units shall therefore identify the component of work in terms of the type of unit, its dimensions and mass.

Mode of description

3.2. To avoid unnecessary length, item descriptions for permanent works shall generally identify the component of the works and not the tasks to be carried out by the contractor, for example

> An item should be described as 'Plain round mild steel bar reinforcement to nominal size 20 mm', not as 'Supply, deliver, cut, bend and fix plain round mild steel bar reinforcement to nominal size 20 mm'.

3.3. Where the work identified by an item is specifically limited, the limitation shall be stated in the item description, for example

> 'Plain round mild steel bar reinforcement to nominal size 20 mm excluding supply and delivery to the Site.'

Item descriptions for work which is divided between two classes require such limitations to be stated, for example

> Item descriptions for miscellaneous metalwork inserts which are to be cast into concrete require appropriate additional description if items are given in both Class G for casting in the inserts and Class N for supplying the inserts.

An item should be described as 'Plain round mild steel bar reinforcement to nominal size 20 mm', not as 'Supply, deliver, cut, bend and fix plain round mild steel bar reinforcement to nominal size 20 mm'.

Separate items

3.4. The work shall be divided into items in the Bill of Quantities so that the component of work which is included in each item does not exhibit more than one feature from each division of any one class of the Work Classification, for example

> One item for precast concrete work shall not include more than one of the types of concrete unit listed in the first division of Class H, neither shall it include different units whose dimensions are not within one of the classifications listed in the second division of Class H, nor shall it include different units whose mass does not lie within one of the ranges listed in the third division of Class H.

Units of measurement

3.5. The unit of measurement for each item shall be that stated for the item in the Work Classification. The unit of measurement stated against a descriptive feature in the Work Classification shall apply to all items to which that descriptive feature applies.

Measurement rules

3.6. Measurement rules in the Work Classification set out the conditions under which work shall be measured and the method by which the quantities shall be computed if other than in accordance with paragraph 5.19.

Definition rules

3.7. Definition rules in the Work Classification define the extent and limits of the class of work represented by a word or expression used in the Work Classification and in a Bill of Quantities prepared in accordance with CESMM4.

Coverage rules

3.8. Coverage rules in the Work Classification provide that the work stated is deemed to be included in the appropriate items to the extent that such work is included in the contract. A coverage rule does not state all the work covered by an item and does not preclude any of the work stated being covered by a Method-Related Charge.

Additional description rules

3.9. Description of an item in addition to that required in accordance with paragraph 3.1 shall be given where required by any provision of section 5 or by any applicable additional description rule in the Work Classification. Where additional description is given, a separate item shall be given for each component of work exhibiting a different additional feature, for example

> Additional description rule A1 of Class H requires that the specification of the concrete in each precast concrete unit shall be stated. Accordingly, this rule also means that separate items shall be given for units cast from concrete of different specifications.

3.10. Where a descriptive feature in the Work Classification identifies a range or group of dimensions and an applicable additional description rule requires the particular dimension to be stated, the range or group of dimensions shall not also be stated, for example

> Additional description rule A2 of Class I requires that the nominal bores of pipes shall be stated in item descriptions. The range of nominal bore taken from the second division of the classification of Class I shall not also be stated.

Applicability of rules

3.11. Rules printed on a right-hand page above a double line apply to all work in the class. Other rules on a right-hand page apply to particular groups of items as shown by the classification table.

Section 4
Coding and numbering of items

Coding

4.1. For convenience of reference each item in the Work Classification has been assigned a code number consisting of a letter and not more than three digits. The letter corresponds to the class in the Work Classification in which the item occurs and the digits give the position of the item in the first, second and third divisions of the class, for example

Code H 1 3 6 identifies an item as

class	H	precast concrete
first division	1	beams
second division	3	length 7–10 m
third division	6	mass 5–10 t

4.2. The symbol * is used in the rules to the Work Classification to indicate all numbers in the appropriate division, for example

H 1 3 * means the group of code numbers from H 1 3 1 to H 1 3 8 inclusive.

Item numbers

4.3. Code numbers may be used to number the items in the Bill of Quantities, the items within the Bill of Quantities being listed in order of ascending code number.

4.4. Code numbers used as item numbers in the Bill of Quantities shall not form part of the item descriptions or be taken into account in the interpretation of the contract.

Coding of unclassified items

4.5. Where a feature of an item is not listed in the Work Classification the digit 9 shall be used in the appropriate positions in the code number.

4.6. Where there is an item to which a division of classification does not apply or for which fewer than three divisions of classification are given the digit 0 shall be used in the appropriate positions in the code number.

Numbering of items with additional description

4.7. Additional description given for an item in accordance with paragraph 3.9 is not represented by the code number. Where code numbers are used as item numbers a suffix number shall be used to distinguish items which have the same code number but different additional description, for example

Additional description rule A1 of Class H requires that additional description be given for precast concrete units stating their position in the works and the specification of the concrete used. If three items are required within one part of the Bill of Quantities to allow for precast concrete beams having the same code but different additional description, the items should be numbered

H 1 3 6. 1
H 1 3 6. 2
H 1 3 6. 3

Section 5
Preparation of the Bill of Quantities

Measurement of completed work

5.1. Appropriate provisions of this section shall also apply to the measurement of completed work.

Sections of the Bill of Quantities

5.2. The Bill of Quantities shall be divided into the following sections.

(*a*) List of principal quantities
(*b*) Preamble
(*c*) Daywork Schedule
(*d*) Work items (grouped into parts)
(*e*) Grand Summary.

List of principal quantities

5.3. A list of the principal components of the works with their approximate estimated quantities shall be given solely to assist tenderers in making a rapid assessment of the general scale and character of the proposed works prior to the examination of the remainder of the Bill of Quantities and the other contractual documents on which their tenders will be based.

Preamble

5.4. The Preamble shall state the methods of measurement other than CESMM4, if any, which have been adopted in the preparation of the Bill of Quantities and are to be used for the measurement of any part of the works. Such methods of measurement shall comprise those adopted and to be used for any work not covered by CESMM4 and any amendments to CESMM4 which have been adopted and are to be used. Amendments comprising abbreviation of CESMM4 are usually necessary for contractor-designed work and work which is intended to involve selection between alternatives at the discretion of the contractor. The extent of the work affected by all amendments to CESMM4 shall be stated in the Preamble.

5.5. Where excavation, boring or driving is included in the work a definition of 'rock' shall be given in the Preamble and this definition shall be used for the purposes of measurement.

5.6. The Method of Measurement is designed to be contract and generally specification neutral. It is therefore necessary to ensure that there is compatibility between the Method of Measurement and the Conditions of Contract used on the project. To achieve this a schedule must be included in the Preamble to the Bill of Quantities indicating the clause in the Conditions of Contract that defines the terminology used in the Method of Measurement.

A schedule is given below of terms found within CESMM4 where such cross reference is required but, if there are amendments made to CESMM4, then further cross references may be required.

If a term is not referred to in a particular contract (e.g. 6.5 Contract price fluctuation), then the schedule should state 'Not applicable' rather than being left blank.

Specific clauses

The following section references in CESMM4 require an appropriate clause and terms in the Conditions of Contract to be drafted to define the term. The schedule shall be replicated in the Preamble to the Bill of Quantities, together with any further required cross references and be completed by the Bill compiler, before the Bill is issued to other parties.

CESMM4 Section Reference		Contract Provision
1.3	Contract administrator	
2.4	Valuation of work executed	
5.1	Measurement of completed work	
5.2	Daywork Schedule	
5.16	Prime cost items	
5.16	Nominated Sub-contractor	
5.18	Provision sums	
6.1	Currency of contract	
6.4	Interim payments	
	Interim certificates	
	Retention moneys	
	Completion	
6.5	Contract price fluctuation	
7.6	Admeasurement	
	Valuing changes	
7.7	Method-Related Charges	
8	Class A Coverage rule C1	
	Class F Definition rule D1	
	Class R Definition rule D1	
†		
†		

† To be completed by the Bill compiler.

Daywork Schedule

5.7. The Daywork Schedule, if any, shall comprise either

(*a*) a list of the various classes of labour, materials and plant for which daywork rates or prices are to be inserted by the tenderer together with a statement of the conditions under which the contractor shall be paid for work executed on a daywork basis, or

(*b*) a statement that the contractor shall be paid for work executed on a daywork basis at rates and prices calculated by adding the percentage additions stated in the standard schedule included in the contract to the rates and prices contained in the aforementioned schedules and by making further adjustments as follows.

Labour	addition/deduction* of — †per cent
Materials	addition/deduction* of — †per cent
Plant	addition/deduction* of — †per cent
Other charges	addition/deduction* of — †per cent

* Appropriate deletion to be made by the contractor when tendering
† Percentage to be inserted by the contractor when tendering

5.8. Provisional Sums for work executed on a daywork basis may be given comprising separate items for labour, materials, plant and other charges. Where a Daywork Schedule of the form stated in sub-paragraph (*b*) of paragraph 5.7 is used each Provisional Sum shall be followed by an item for the adjustment referred to in that sub-paragraph. The price inserted against each such item shall be calculated by applying the percentage addition or deduction inserted by the contractor in the Daywork Schedule to the amount of the associated Provisional Sum.

Work items

Division of the Bill of Quantities into parts

5.9. The items in the Bill of Quantities which are to be priced and to contribute to the Tender Total may be arranged into numbered parts to distinguish between those parts of the work of which the nature, location, access, limitation on sequence or timing or any other special characteristic is thought likely to give rise to different methods of construction or considerations of cost. General items (Class A) may be grouped as a separate part of the Bill of Quantities. Items in each part shall be arranged in the general order of the Work Classification.

Headings and sub-headings

5.10. Each part of the Bill of Quantities shall be given a heading and groups of items within each part be given sub-headings. Headings and sub-headings shall be read as part of the item descriptions to which they apply. A line shall be drawn across the item description column below the last item to which each heading or sub-heading applies. Headings and sub-headings shall be repeated at the start of each new page which lists items to which they apply.

Extent of itemization and description

5.11. All work shall be itemized and the items shall be described in accordance with the Work Classification, but further itemization and additional description may be provided if the nature, location, importance or any other special characteristic of the work is thought likely to give rise to special methods of construction or considerations of cost.

Descriptions

5.12. Descriptions shall identify the work covered by the respective items, but the exact nature and extent of the work is to be ascertained from the Drawings, Specification and Conditions of Contract, as the case may be, read in conjunction with the Work Classification.

5.13. Any detail of description required to be given in accordance with the Work Classification may be omitted from an item description provided that a reference is given in its place which identifies precisely where the omitted information may be found on a drawing or in the Specification.

5.14. Where an item description compiled in accordance with the Work Classification would be insufficient to identify clearly the particular work covered by the item additional description shall be given to identify the work by reference to its location or other physical features shown on the Drawings or described in the Specification.

Ranges of dimensions

5.15. Where all the components of work included in an item are of one dimension within a range given in the Work Classification that dimension may be stated in the item description in place of the range of dimensions given.

Prime Cost Items

5.16. The estimated price of work to be carried out by a Nominated Sub-contractor shall be given in the Bill of Quantities as a Prime Cost Item. Each Prime Cost Item shall be followed by

(*a*) an item for a sum for labours in connection therewith which, in the absence of any express provision in the contract to the contrary, shall include *only*
 (i) in any case in which the Nominated Sub-contractor is to carry out work on the Site for allowing him to use temporary roads, scaffolding, hoists, messrooms, sanitary accommodation and welfare facilities which are provided by the contractor for his own use and for providing space for office accommodation and storage of plant and materials, for disposing of rubbish and for providing light and water for the work of the Nominated Sub-contractor, and
 (ii) in any case in which the Nominated Sub-contractor is not to carry out work on the Site for unloading, storing and hoisting materials supplied by him and returning packing materials, and

(*b*) an item expressed as a percentage of the price of the Prime Cost Item in respect of all other charges and profit.

5.17. Where any goods, materials or services supplied by a Nominated Sub-contractor are to be used by the contractor in connection with any item, reference shall be made in the description of that item, or in the appropriate heading or sub-heading, to the Prime Cost Item under which the goods or materials or services are to be supplied.

Provisional Sums

5.18. Provision for contingencies shall be made by giving Provisional Sums in the Bill of Quantities and not by increasing the quantities beyond those of the work expected to be required. Provisional Sums for Defined Work shall be included where work is known to be required but the scope of the work cannot be completely designed but the scope can be defined. A Provisional Sum for a general contingency allowance shall be given in the Grand Summary in accordance with paragraph 5.26.

Quantities

5.19. The quantities shall be computed net using dimensions from the Drawings, unless directed otherwise by a measurement rule in CESMM4 or by the Contract, and no allowance shall be made for bulking, shrinkage or waste. Quantities may be rounded up or down where appropriate. Fractional quantities are not generally necessary and should not be given to more than one place of decimals.

Units of measurement

5.20. The following units of measurement and abbreviations shall be used.

Unit	*Abbreviation*
Millimetre	mm
Metre	m
Square millimetre	mm^2 or mm2
Square metre	m^2 or m2
Hectare	ha
Cubic metre	m^3 or m3
Kilogramme	kg
Tonne	t
Sum	sum
Number	nr
Hour	h
Week	wk

Work affected by water

5.21. Where an existing body of open water (other than groundwater) such as a river, stream, canal, lake or body of tidal water is either on the Site or bounds the Site, each body of water shall be identified in the Preamble to the Bill of Quantities. A reference shall also be given to a drawing indicating the boundaries and surface level of each body of water or, where the boundaries and surface levels fluctuate, their anticipated ranges of fluctuation.

Ground and excavation levels

5.22. The Commencing Surface shall be identified in the description of each item for work involving excavation, boring or driving for which the Commencing Surface is not the Original Surface. The Excavated Surface shall be identified in the description of each item for work involving excavation for which the Excavated Surface is not the Final Surface. The depths of excavation stated in accordance with the Work Classification shall be measured from the Commencing Surface to the Excavated Surface.

5.23. Provision shall be made for the amounts inserted on each page to be totalled and carried to a summary of each part of the Bill of Quantities and for the total of each Part Summary to be carried to the Grand Summary.

Grand Summary

5.24. The Grand Summary shall contain a tabulation of the parts of the Bill of Quantities with provision for insertion of the total of the amounts brought forward from the Part Summaries.

General Contingency Allowance

5.25. An item for a general contingency (the General Contingency Allowance), if instructed by the employer, shall be given in the Grand Summary following the total of the amounts brought forward from the Part Summaries.

Adjustment Item

5.26. An item described as the Adjustment Item shall be given in the Grand Summary following the total of the amounts brought forward from the Part Summaries and the General Contingency Allowance, if any (*see* paragraphs 6.3 and 6.4).

Total of the Priced Bill of Quantities

5.27. Provision shall be made for insertion of the total of the amounts brought forward from the Part Summaries, the amount of the General Contingency Allowance, if any, and the amount of the Adjustment Item.

Section 6
Completion, pricing and use of the Bill of Quantities

Insertion of rates and prices

6.1. Rates and prices shall be inserted in the rate column of the Bill of Quantities in the currency of the contract.

Parts to be totalled

6.2. Each part of the Bill of Quantities shall be totalled and the totals carried to the Grand Summary.

Adjustment Item

6.3. A tenderer may insert a lump sum addition or deduction against the Adjustment Item given in the Grand Summary in adjustment of the total of the priced Bill of Quantities.

6.4. For the purposes of interim payments additions or deductions on account of the amount, if any, of the Adjustment Item shall be made by instalments in interim certificates in the proportion that the amount bears to the total of the priced Bill of Quantities before the addition or deduction of the amount of the Adjustment Item and a statement to this effect shall appear in the Preamble to the Bill of Quantities. Such interim additions or deductions shall be made before deduction of the retention moneys if any, and shall not exceed in the aggregate the amount of the Adjustment Item. If by the date of the completion of the works any balance of the amount of the Adjustment Item is outstanding it shall be added to or deducted from the moneys then due.

6.5. For the purposes of a contract price fluctuations clause, if applicable, account shall be taken of any addition to or deduction from the amounts due to the contractor in respect of the Adjustment Item.

Section 7
Method-related charges

Definitions

7.1. For the purposes of this section the following words and expressions shall have the meanings hereby assigned to them.

(*a*) 'Method-Related Charge' means the sum for an item inserted in the Bill of Quantities by a tenderer in accordance with paragraph 7.2.
(*b*) 'Time-Related Charge' means a Method-Related Charge for work the cost of which is to be considered as proportional to the length of time taken to execute the work.
(*c*) 'Fixed Charge' means a Method-Related Charge which is not a Time-Related Charge.

Insertion by a tenderer

7.2. A tenderer may insert in the Bill of Quantities such items for Method-Related Charges as he may decide to cover items of work relating to his intended method of executing the Works, the costs of which are not to be considered as proportional to the quantities of the other items and for which he has not allowed in the rates and prices for the other items.

Itemization

7.3. Where possible the itemization of Method-Related Charges should follow the order of classification and the other requirements set out in class A of the Work Classification, distinguishing between Time-Related Charges and Fixed Charges. Method-Related Charges may be inserted to cover items of work other than those set out in class A.

Description

7.4. Each item for a Method-Related Charge inserted in the Bill of Quantities shall be fully described so as to define precisely the extent of the work covered and to identify the resources to be used and the particular items of Permanent Works or Temporary Works, if any, to which the item relates.

Contractor not bound to adopt method

7.5. The insertion by the contractor of an item for a Method-Related Charge in the Bill of Quantities when tendering shall not bind him to adopt the method stated in the item description in executing the Works.

Charges not to be measured

7.6. Method-Related Charges shall not be subject to admeasurement but shall be deemed to be prices for the purposes of valuing changes to the works and revisions to rates as a consequence of a change in quantity arising from admeasurement.

Payment

7.7. Method-Related Charges shall be certified and paid in accordance with the contract and a statement to this effect shall appear in the Preamble to the Bill of Quantities.

Payment when method not adopted

7.8. In the event of the satisfactory execution of any part of the works which has been the subject of an item for a Method-Related Charge using, whether in whole or in part, a method other than that described in the item the contractor shall nevertheless be entitled to payment of the Method-Related Charge or the balance thereof, as the case may be, by such installments at such times and upon such events as may from time to time be agreed between the contract administrator and the contractor. In default of such agreement the Method-Related Charge, or the balance then unpaid, shall be treated as if it were an addition to the Adjustment Item referred to in paragraphs 6.3, 6.4 and 6.5 and allowed to the contractor by way of payments in interim certificates. The amount of a Method-Related Charge shall be neither increased nor decreased by reason only of any change in method made by the contractor, unless such change has been ordered by the contract administrator, in which case the provisions of the contract in valuing changes shall apply.

Section 8
Work classification

CLASS A: GENERAL ITEMS

Includes: **General obligations, site services and facilities, Temporary Works, testing of materials and work, Provisional Sums and Prime Cost Items**
Items to cover elements of the cost of the work which are not to be considered as proportional to the quantities of the Permanent Works

FIRST DIVISION	SECOND DIVISION	THIRD DIVISION
1 Contractual requirements	**1** Performance bond **2** Insurances **3** Parent company guarantee	
2 Specified requirements	**1** Accommodation for the contract administrators	**1** Offices **2** Laboratories **3** Cabins
	2 Services for the contract administrators	**1** Transport vehicles **2** Telephones
	3 Equipment for use by the contract administrators	**1** Office equipment **2** Laboratory equipment **3** Surveying equipment
	4 Attendance upon the contract administrators	**1** Drivers **2** Chainmen **3** Laboratory assistants
	5 Testing of materials **6** Testing of the Works	
	7 Temporary Works	**1** Traffic diversions **2** Traffic regulation **3** Access roads **4** Bridges **5** Cofferdams **6** Pumping **7** De-watering **8** Compressed air for tunnelling
3 Method-Related Charges	**1** Accommodation and buildings	**1** Offices **2** Laboratories **3** Cabins **4** Stores **5** Canteens and messrooms
	2 Services	**1** Electricity **2** Water **3** Security **4** Hoardings **5** Site transport **6** Personnel transport **7** Welfare
	3 Plant	**1** Cranes **2** Transport **3** Earthmoving **4** Compaction **5** Concrete mixing **6** Concrete transport **7** Pile driving **8** Pile boring
	4 Plant	**1** Pipelaying **2** Paving **3** Tunnelling **4** Crushing and screening **5** Boring and drilling

(Continued)

MEASUREMENT RULES	DEFINITION RULES	COVERAGE RULES	ADDITIONAL DESCRIPTION RULES
M1 The unit of measurement for general items may be the sum, except where another unit of measurement is used in accordance with rule M2.			
M2 An item may be given for a *Parent Company Guarantee* where it is envisaged a tenderer may be a subsidiary of a larger company or group.		**C1** Items for *insurances* classed as *contractual requirements*	
M3 A quantity shall be given against all items for *specified requirements* of which the value is to be ascertained and determined by admeasurement. A unit of measurement shall be stated for each such item. **M4** Items shall be given in this class for all *testing* for which items are not given separately as set out in other classes.	**D1** All work other than the permanent works which is expressly stated in the contract to be carried out by the contractor and of which the nature and extent is expressly stated in the contract shall be classed as *specified requirements*.		**A1** Item descriptions for work classed as *specified requirements* which is to be carried out after the date for completion shall so state. **A2** Item descriptions for work classed as *specified requirements* shall distinguish between the establishment and removal of services or facilities and their continuing operation or maintenance. **A3** Item descriptions for *testing of materials* and *testing of the works* shall include particulars of samples and of methods of testing.
M5 Items for *Method-Related Charges,* if any, shall be inserted by the tenderer in accordance with section 7.			**A4** Item descriptions for *Method-Related Charges* shall distinguish between Fixed and Time-Related Charges.

CLASS A

FIRST DIVISION	SECOND DIVISION	THIRD DIVISION
3 Method-Related Charges (*continued*)	**5** Temporary Works	**1** Traffic diversions **2** Traffic regulation **3** Access roads **4** Bridges **5** Cofferdams **6** Pumping **7** De-watering **8** Compressed air for tunnelling
	6 Temporary Works	**1** Access scaffolding **2** Support scaffolding and propping **3** Piling **4** Formwork **5** Shafts and pits **6** Hardstandings
	7 Supervision and labour	**1** Supervision **2** Administration **3** Labour teams
4 Provisional Sums	**1** Daywork	**1** Labour **2** Percentage adjustment to Provisional Sum for Daywork labour **3** Materials **4** Percentage adjustment to Provisional Sum for Daywork materials **5** Plant **6** Percentage adjustment to Provisional Sum for Daywork plant **7** Other charges **8** Percentage adjustment to Provisional Sum for Daywork other charges
	2 Provisional Sums – Defined Work	
5 Nominated Sub-contracts which include work on the Site **6** Nominated Sub-contracts which do not include work on the Site	**1** Prime Cost Item **2** Labours **3** Special labours **4** Other charges and profit	

NOTE

Method-Related Charges may be inserted by the tenderer in accordance with paragraph 7.2 for insurances additional to those classed as contractual requirements.

MEASUREMENT RULES	DEFINITION RULES	COVERAGE RULES	ADDITIONAL DESCRIPTION RULES
M6 Items for *percentage adjustment to Provisional Sums for Daywork* shall be given only where a Daywork Schedule in accordance with alternative form (b) of paragraph 5.6 is given in the Bill of Quantities. Adjustments shall be inserted against such items to correspond with the adjustments, if any, inserted by the tenderer in the Daywork Schedule.			
M7 *Provisional Sums – Defined Work* shall be given where the scope of the work cannot be completely designed but where the scope can be defined in rule C2.	**D2** *Provisional Sums – Defined Work* where the scope of the work cannot be completely designed but where the scope can be defined	**C2** Where *Provisional Sums – Defined Work* are included the Tenderer is deemed to have made due allowances for the programming, planning and pricing of the Works within his General Items.	**A5** Item descriptions for *Provisional Sums – Defined Work* shall state the nature of the work, indicative quantities of the scope and extent of the work.
M8 Each *Prime Cost Item* shall be followed by an item for labours and an item for other charges and profit in accordance with paragraph 5.15. Where labours other than or in addition to those stated in sub-paragraph (a) of paragraph 5.15 are to be provided the item for labours shall be designated as for special labours.			**A6** Item descriptions for *Prime Cost Items* shall identify the work included. **A7** The labours shall be stated in item descriptions for *special labours*.

CLASS B: GROUND INVESTIGATION

Includes: **Trial pits and trenches, boreholes, samples, site and laboratory tests, instrumental observations and professional services in connection with ground investigation**
Excludes: **Excavation not carried out for the purpose of ground investigation (included in class E) Boring for piling (included in classes P and Q)**

FIRST DIVISION	SECOND DIVISION		THIRD DIVISION
1 Trial pits and trenches	**1** Number in material other than rock **2** Number in material which includes rock	nr nr	**1** Maximum depth: not exceeding 1 m **2** 1–2 m **3** 2–3 m **4** 3–5 m **5** 5–10 m **6** 10–15 m **7** 15–20 m **8** stated exceeding 20 m
	3 Depth in material other than rock **4** Depth in rock **5** Depth supported **6** Depth backfilled, material stated	m m m m	
	7 Removal of obstructions	h	
	8 Pumping at a stated minimum extraction rate	h	
2 Light cable percussion boreholes	**1** Number	nr	
	3 Depth	m	**1** In holes of maximum depth: not exceeding 5 m **2** 5–10 m **3** 10–20 m **4** 20–30 m **5** 30–40 m **6** stated exceeding 40 m
	6 Depth backfilled, material stated	m	
	7 Chiselling to prove rock or to penetrate obstructions	h	
3 Rotary drilled boreholes	**1** Number	nr	
	3 Depth without core recovery **4** Depth with core recovery	m m	**1** In holes of maximum depth: not exceeding 5 m **2** 5–10 m **3** 10–20 m **4** 20–30 m **5** 30–40 m **6** stated exceeding 40 m
	5 Depth cased **6** Depth backfilled, material stated **7** Core boxes, length of core stated	m m nr	

MEASUREMENT RULES	DEFINITION RULES	COVERAGE RULES	ADDITIONAL DESCRIPTION RULES
		C1 Items for ground investigation shall be deemed to include preparation and submission of records and results. **C2** Items for ground investigation shall be deemed to include disposal of excavated material and removal of dead services.	
			A1 Item descriptions for the number and depth of *trial pits and trenches* shall state the minimum plan area at the bottom of the pit or trench or, where the work is undertaken to locate services, the maximum length of the trench. **A2** Item descriptions for the number and depth of *trial pits and trenches* (B 1 1–4 *) shall identify separately those which are expressly required to be excavated by hand.
			A3 Item descriptions for *pumping* shall state any special de-watering methods which are expressly required.
			A4 Item descriptions for the number and depth of *light cable percussion boreholes* shall state the nominal diameter of the bases of the boreholes.
		C3 Items for the *depth* of light cable percussion boreholes shall be deemed to include casings.	
M1 *Chiselling to prove rock or to penetrate obstructions* shall be measured only where it is expressly required.			
	D1 *Rotary drilled boreholes* shall be classified as such whether the boring is carried out by a rig or by an attachment. **D2** *Core boxes* shall be deemed to become the property of the Employer unless otherwise stated. **D3** The maximum depth used for classification of *rotary drilled boreholes* which are continuations of light cable percussion boreholes shall be measured from the Commencing Surface of the light cable percussion borehole.		**A5** Item descriptions for *rotary drilled boreholes* shall state the nominal minimum core diameter. **A6** Item descriptions for the *number* of rotary drilled boreholes shall state those which are continuations of light cable percussion boreholes. **A7** Item descriptions for *rotary drilled boreholes* which are inclined shall state the angle of inclination.

CLASS B

FIRST DIVISION		SECOND DIVISION	THIRD DIVISION	
4 Samples	nr	**1** From the surface or from trial pits and trenches	**1** Undisturbed soft material **2** Disturbed soft material **3** Rock **4** Groundwater	
		2 From boreholes	**1** Open tube **2** Disturbed **3** Groundwater **4** Stationary piston **5** Swedish foil **6** Delft **7** Bishop sand	
5 Site tests and observations		**1**	**1** Permeability	h
			2 Groundwater level	nr
			3 Standard penetration	nr
			4 Penetration vane **5** Vane in borehole **6** Pressure meter	nr nr nr
		2	**1** Plate bearing	nr
			2 Self-boring pressure meter **3** California bearing ratio	nr nr
			4 Static cone sounding	nr
			5 In situ density **6** Mackintosh probe	nr nr
			7 Hand auger borehole	nr
6 Instrumental observations		**1** Pressure head	**1** Standpipes **2** Piezometers **3** Install covers **4** Readings	m m nr nr
		2 Inclinometers	**1** Installations **4** Readings	m nr
		3 Settlement gauges **4** Resistivity **5** Seismic **6** Magnetometer **7** Self-potential **8** Gravimetric	**1** Installations **4** Readings	nr nr

MEASUREMENT RULES	DEFINITION RULES	COVERAGE RULES	ADDITIONAL DESCRIPTION RULES
			A8 Item descriptions for *samples* shall state their size, type and class.
			A9 Item descriptions for *permeability* tests shall state the type and give particulars of the tests.
			A10 Item descriptions for *groundwater level* observations shall state when the measurements are to be taken.
			A11 Item descriptions for *standard penetration* tests shall state whether they are in light cable percussion boreholes or rotary drilled boreholes.
			A12 Item descriptions for *plate bearing* tests shall state whether they are in pits and trenches or boreholes or at the surface.
			A13 Item descriptions for *static cone sounding* tests shall state the maximum depth of the cone and, where electric cones are used, the maximum capacity of the machine.
			A14 Item descriptions for *hand auger borehole* tests and observations shall state the minimum diameter and the maximum depth of the boreholes.
			A15 Item descriptions for *instrumental observations* shall state details of the type of observations and the type of protective fences.
		C4 Items for *inclinometers* and *settlement gauges* shall be deemed to include provision of special boreholes.	**A16** Item descriptions for *inclinometers* and *settlement gauges* shall state whether the instruments are in special boreholes.

CLASS B

FIRST DIVISION		SECOND DIVISION	THIRD DIVISION
7 Laboratory tests	nr	**1** Classification	**1** Moisture content **2** Atterberg limits **3** Specific gravity **4** Particle size analysis by sieve **5** Particle size analysis by pipette or hydrometer **6** Frost susceptibility
		2 Chemical content	**1** Organic matter **2** Sulphate **3** pH value **4** Contaminants
		3 Compaction	**1** Standard **2** Heavy **3** Vibratory
		4 Consolidation	**1** Oedometer cell **2** Triaxial cell **3** Rowe cell
		5 Permeability	**1** Constant head **2** Falling head
		6 Soil strength	**1** Quick undrained triaxial **2** Consolidated undrained triaxial, with pore water pressure measurement **3** Consolidated drained triaxial, with volume change measurement **4** Shearbox: peak only **5** peak and residual **6** residual only **7** ring shear **8** California bearing ratio
		7 Rock strength	**1** Unconfined compressive strength of core samples **2** Consolidated drained triaxial, with volume change measurement **3** Brazilian **4** Ring shear **5** Point load

CLASS B

MEASUREMENT RULES	DEFINITION RULES	COVERAGE RULES	ADDITIONAL DESCRIPTION RULES
			A17 Item descriptions for tests for *contaminants* shall state the standards required and the contaminants to be analysed.
			A18 Item descriptions for *triaxial cell* and *Rowe cell* tests shall state the number of increments and the effective pressures.
			A19 Item descriptions for *quick undrained triaxial* tests shall state the diameter and whether a single, multistage or set of three specimens is required. **A20** Item descriptions for *consolidated triaxial* tests shall state the diameter and the effective pressures. Multistage tests shall be identified in item descriptions. **A21** Item descriptions for *shearbox* tests shall state the normal pressures and the size of the shearbox. **A22** Item descriptions for *California bearing ratio* tests shall state the compactive effort, surcharge and whether soaking is required.
			A23 Item descriptions for *unconfined compressive strength* tests shall state the diameter and the height of the samples. **A24** Item descriptions for *consolidated drained triaxial* tests shall state the diameter and the height of the samples and the effective pressure. **A25** Item descriptions for *Brazilian* tests shall state the diameter and the length of the samples. **A26** Item descriptions for *ring shear* tests shall state the normal pressure and the diameter of the samples. **A27** Item descriptions for *point load* tests shall state the type of test required and the minimum dimensions of the samples.

CLASS B

FIRST DIVISION	SECOND DIVISION		THIRD DIVISION
8 Professional services	**1** Technician **2** Technician engineer	h h	
	3 Engineer or geologist	h	**1** Graduate **2** Chartered **3** Principal or consultant
	4 Visits to the Site **5** Overnight stays in connection with visits to the Site	nr nr	

MEASUREMENT RULES	DEFINITION RULES	COVERAGE RULES	ADDITIONAL DESCRIPTION RULES
M2 *Professional services* shall be measured only where they are expressly required for analysis of records and results. **M3** The hours measured shall be working hours and shall exclude hours occupied in travel, meals, etc.		**C5** Items for *professional services* shall be deemed to include preparation and submission of reports and keeping records of time spent. **C6** Items for *visits to the Site* and *overnight stays in connection with visits to the Site* shall be deemed to include travelling, meals, accommodation and other incidental expenses.	

CLASS C: GEOTECHNICAL AND OTHER SPECIALIST PROCESSES

Includes: **Geotechnical processes for altering the properties of soils and rocks**
Other specialist processes as listed
Excludes: **Compaction (included in class E)**
Grouting carried out from within tunnels, shafts and other subterranean cavities (included in class T)
Grouting carried out from within sewers (included in class Y)

FIRST DIVISION	SECOND DIVISION	THIRD DIVISION
1 Drilling for grout holes through material other than rock or artificial hard material **2** Drilling for grout holes through rock or artificial hard material **3** Driving injection pipes for grout holes	**1** Vertically downwards **2** Downwards at an angle 0°–45° to the vertical **3** Horizontally or downwards at an angle less than 45° to the horizontal **4** Upwards at an angle 0°–45° to the horizontal **5** Upwards at an angle less than 45° to the vertical	**1** In holes of depth: not exceeding 5 m — m **2** 5–10 m — m **3** 10–20 m — m **4** 20–30 m — m **5** 30–40 m — m **6** stated exceeding 40 m — m
4 Grout holes materials and injection	**1** Number of holes — nr **2** Number of stages — nr **3** Single water pressure tests — nr **4** Multiple water pressure tests — nr **5** Materials — t	**1** Cement **2** Pulverized fuel ash **3** Sand **4** Pea gravel **5** Bentonite **6** Chemicals
	6 Injection	**1** Number of injections — nr **2** Neat cement grout — t **3** Cement and stated filler grout — t **4** Chemical grout — t **5** Other stated grout — t **6** Single packer settings — nr **7** Double packer settings — nr
5 Diaphragm walls	**1** Excavation in material other than rock or artificial hard material — m³ **2** Excavation in rock — m³ **3** Excavation in artificial hard material — m³	**1** Maximum depth: not exceeding 5 m **2** 5–10 m **3** 10–15 m **4** 15–20 m **5** 20–25 m **6** 25–30 m **7** stated exceeding 30 m
	4 Concrete — m³	
	5 Plain round steel bar reinforcement — t **6** Deformed high yield steel bar reinforcement — t	**1** Nominal size: 6 mm **2** 8 mm **3** 10 mm **4** 12 mm **5** 16 mm **6** 20 mm **7** 25 mm **8** 32 mm or greater
	7 Waterproofed joints — sum **8** Guide walls — m	

MEASUREMENT RULES	DEFINITION RULES	COVERAGE RULES	ADDITIONAL DESCRIPTION RULES
M1 The Commencing Surface adopted in the preparation of the Bill of Quantities shall be adopted for the measurement of the completed work. **M2** The depths of *grout holes*, holes for *ground reinforcement* and *drains* shall be measured along the holes irrespective of inclination.	**D1** Drilling and excavation for work in this class shall be deemed to be in *material other than rock or artificial hard material* unless otherwise stated in item descriptions.	**C1** Items for work in this class shall be deemed to include disposal of excavated material and removal of dead services.	
M3 Drilling through previously grouted holes in the course of stage grouting shall not be measured. Where holes are expressly required to be extended, the number of holes shall be measured and drilling through previously grouted holes shall be measured as *drilling through rock or artificial hard material*. **M4** The *number of stages* measured shall be the total number of grouting stages expressly required.			**A1** The diameters of holes shall be stated in item descriptions for *drilling* and *driving for grout holes* and *grout holes*.
M5 The mass of *grout materials* measured shall not include the mass of mixing water.			**A2** The type of materials shall be stated in item descriptions for *grout materials*.
M6 The *number of injections* measured shall be the total number of injections expressly required. **M7** The mass of *grout injection* measured shall not include the mass of mixing water.			**A3** Item descriptions for the *number of injections* shall identify those which are in stages distinguishing between those which are in ascending and descending stages.
M8 Formwork for voids, rebates and fillets in *diaphragm walls* shall be classed as concrete ancillaries (class G). **M9** The depths of *concrete* in diaphragm walls shall be measured from the cut-off levels expressly required. The volume of concrete shall be calculated as set out in rules M1 and M2 in class F. **M10** The mass measured for *reinforcement* in diaphragm walls shall include that of stiffening, lifting and supporting steel cast in. **M11** The mass of steel *reinforcement* shall be taken as 0.785 kg/m per 100 mm^2 of cross-section (7.85 t/m^3). The mass of other reinforcing materials shall be taken as stated in the Contract. **M12** *Guide walls* shall be measured each side of the diaphragm wall.	**D2** *Diaphragm walls* are walls constructed using bentonite slurry or other support fluids. **D3** The nominal size stated in item descriptions for *bar reinforcement* in diaphragm walls shall be the cross-sectional size.	**C2** Items for *excavation* for diaphragm walls shall be deemed to include preparation and upholding sides of excavation. **C3** Items for *concrete* in diaphragm walls shall be deemed to include trimming the faces of diaphragm walls and preparing their tops to receive other work. **C4** Items for *reinforcement* in diaphragm walls shall be deemed to include supporting reinforcement and preparing protruding reinforcement to receive other work.	**A4** The thicknesses of diaphragm walls shall be stated in item descriptions for *excavation* and *concrete* for diaphragm walls. **A5** The nature of the material shall be stated in item descriptions for *excavation in artificial hard material*. **A6** The mix specifications or strengths shall be stated in item descriptions for *concrete* in diaphragm walls.

CLASS C

FIRST DIVISION	SECOND DIVISION	THIRD DIVISION
6 Ground reinforcement	**1** Number in material other than rock or artificial hard material to a stated maximum depth nr **2** Total length of tendons in material other than rock or artificial hard material m **3** Number in material which includes rock or artificial hard material to a stated maximum depth nr **4** Total length of tendons in material which includes rock or artificial hard material m	**1** Temporary **2** Temporary with single corrosion protection **3** Temporary with double corrosion protection **4** Permanent **5** Permanent with single corrosion protection **6** Permanent with double corrosion protection
7 Sand, band and wick drains	**1** Number of drains nr **2** Number of predrilled holes nr **3** Depth of overlying material m **4** Depth of drains of maximum depth: not exceeding 10 m m **5** 10–15 m m **6** 15–20 m m **7** 20–25 m m **8** stated exceeding 25 m m	**1** Cross-sectional dimension: not exceeding 100 mm **2** 100–200 mm **3** 200–300 mm **4** 300–400 mm **5** 400–500 mm **6** stated exceeding 500 mm
8 Vibro floation	**1** Number of probes nr **2** Depth of probe	**1** Depth not exceeding 5 m m **2** Depth: 5–10 m m **3** Depth: 10–20 m m **4** Depth: 20–30 m m **5** Depth: 30–40 m m **6** Depth: stated exceeding 40 m m

MEASUREMENT RULES	DEFINITION RULES	COVERAGE RULES	ADDITIONAL DESCRIPTION RULES
M13 The *lengths of tendons* for ground anchorages shall be measured between the outer ends of anchorages	**D4** Ground anchorages and soil nailing shall be classed as *ground reinforcement.*		**A7** The composition, location and working load and details of water and grout testing, pregrouting and grouting shall be stated in item descriptions for ground anchorages. **A8** The composition, location and details of testing shall be stated in item descriptions for soil mailing.
M14 The number of *predrilled holes* measured for sand, band and wick drains shall be the number expressly required to be predrilled through overlying material.			**A9** Sand drains, band drains and wick drains shall be separately identified in item descriptions and the materials of which they are composed stated.

CLASS D: DEMOLITION AND SITE CLEARANCE

Includes: **Demolition and removal of natural and artificial articles, objects and obstructions which are above the Original Surface**

Excludes: **Removal of articles, objects, obstructions and materials (other than tree roots) at or below the Original Surface (included in classes C, E, I, J, K, L, R, T, X and Y)**

FIRST DIVISION		SECOND DIVISION		THIRD DIVISION
1 Site clearance		**1** General clearance **2** Invasive plant species	ha m^2	
2 Trees	nr	**1** Girth: 500 mm–2 m **2** exceeding 2 m		
3 Stumps	nr	**1** Diameter: less than 1 m **2** exceeding 1 m		
4 Buildings **5** Other structures	sum sum	**1** Brickwork **2** Concrete **3** Masonry **4** Metal **5** Timber **6** No predominant material		**1** Volume: not exceeding 50 m^3 **2** 50–100 m^3 **3** 100–250 m^3 **4** 250–500 m^3 **5** 500–1000 m^3 **6** 1000–2500 m^3 **7** 2500–5000 m^3 **8** stated exceeding 5000 m^3
6 Pipelines	m	**1** Nominal bore: 100–300 mm **2** 300–500 mm **3** exceeding 500 mm		

MEASUREMENT RULES	DEFINITION RULES	COVERAGE RULES	ADDITIONAL DESCRIPTION RULES
		C1 Items for demolition and site clearance shall be deemed to include disposal of the materials arising.	**A1** Item descriptions for work from which the materials arising remain the property of the Employer shall so state.
	D1 *General clearance* shall include the demolition and removal of all articles, objects and obstructions which are expressly required to be cleared, except those for which separate items are given as set out in this class. **D2** *Invasive plant species* shall include any plant whose control is governed by legislation and which is treated by herbicidal or chemical process.	**C2** Items for *general clearance* which include the removal of hedges shall be deemed to include the removal of hedge stumps of any diameter where these are also required to be removed. **C3** Items for *invasive plant species* shall include for the disposal of any vegetable matter remaining after treatment.	**A2** Item descriptions for *general clearance* shall identify the area included unless it is the total area of the Site. **A3** Item descriptions for *invasive plant species* shall state the method of treatment and type of vegetation. **A4** Where holes left by stump removal are to be backfilled, item descriptions for *general clearance, trees* and *stumps* shall state the nature of the backfilling material.
	D3 Girths of *trees* shall be measured 1 m above ground level.	**C4** Items for clearance of *trees* shall be deemed to include removal of the stumps where they are also required to be removed.	
	D4 The volume used in the classification of *buildings* and *other structures* shall be their approximate volume occupied, excluding any volume below the Original Surface.		**A5** *Buildings* and *other structures* shall be identified in item descriptions.
M1 *Pipelines* within buildings and other structures shall be measured only where their nominal bore exceeds 300 mm.		**C5** Items for demolition of *pipelines* shall be deemed to include demolition and removal of supports.	

CLASS E: EARTHWORKS

Includes: **Excavation, dredging, filling, compaction, disposal and landscaping**
Excludes: **Excavation for:**
ground investigation (included in class B)
diaphragm walls and ground anchorages (included in class C)
pipes and sewers, manholes, trenches and ditches, pipe headings, thrust boring and pipe jacking, and pipe jointing (included in classes I, J, K, L and Y)
piles (included in classes P and Q)
foundations for traffic signs (included in class R)
tunnels, shafts, headings and other subterranean cavities (included in class T)
foundations for fences and gates (included in class X)
Reinstatement following pipe laying (included in class K)

FIRST DIVISION		SECOND DIVISION	THIRD DIVISION
1 Excavation by dredging	m^3	**1** Topsoil	
2 Excavation for cuttings	m^3	**2** Material other than topsoil, rock or artificial hard material	
3 Excavation for foundations	m^3	**3** Rock	**1** Maximum depth: not exceeding 0.25 m
4 General excavation	m^3	**4** Stated artificial hard material exposed at the Commencing Surface	**2** 0.25–0.5 m
		5 Stated artificial hard material not exposed at the Commencing Surface	**3** 0.5–1 m
		6 Controlled and hazardous material exposed at the Commencing Surface	**4** 1–2 m
		7 Controlled and hazardous material not exposed at the Commencing Surface	**5** 2–5 m
			6 5–10 m
			7 10–15 m
			8 stated exceeding 15 m

CLASS E

MEASUREMENT RULES	DEFINITION RULES	COVERAGE RULES	ADDITIONAL DESCRIPTION RULES
M1 In accordance with paragraph 5.18 the quantities of earthworks shall be computed net using dimensions from the Drawings with no allowance for bulking, shrinkage or waste. Where boundaries between different materials are not shown on the Drawings, measurement shall be made on the Site.			
M2 The Commencing Surface adopted in the preparation of the Bill of Quantities shall be adopted for the measurement of the completed work. **M3** Excavation classed as *by dredging* in the Bill of Quantities shall be measured as by dredging irrespective of the method of excavation adopted by the Contractor. **M4** Measurement of *excavation by dredging* shall be made from soundings unless otherwise stated. **M5** An item shall be given for each separate stage of excavation where separate stages in the conduct of the Works are expressly required. **M6** The volume measured for the excavation of a structure or foundation shall be the volume which is to be either occupied by or vertically above any part of the structure or foundation. **M7** The volume measured for excavation below a body of open water shall be the volume below water when the water surface is at the level (or the higher level of fluctuation if applicable) shown on the drawing to which reference is given in the Preamble in accordance with paragraph 5.20. **M8** An isolated volume of *artificial hard material* or *rock* occurring within other material to be excavated shall not be measured separately unless its volume exceeds 1 m^3 except that the minimum volume shall be 0.25 m^3 where the net width of excavation is less than 2 m. **M9** The volume measured for excavation within borrow pits shall be the net volume measured for filling.	**D1** Excavated material shall be deemed to be *material other than topsoil, rock or artificial hard material* unless otherwise stated in item descriptions. **D2** Excavation in or under an embankment, executed prior to placing of fill, shall be classed as *excavation for cuttings*. **D3** Excavation from within borrow pits shall be classed as *general excavation*. **D4** *Controlled and hazardous materials* shall include any material whose excavation and disposal is governed by legislation including invasive plant species.	**C1** Items for excavation shall be deemed to include upholding sides of excavation, additional excavation to provide working space and removal of dead services. **C2** Item descriptions for excavation shall be deemed to include removal of existing pipes of any material or diameter. **C3** Items for excavation within borrow pits shall be deemed to include removal and replacement of overburden and unsuitable material.	**A1** The location and limits of *excavation by dredging* shall be stated in item descriptions where its extent would otherwise be uncertain. **A2** Item descriptions for excavation below a body of open water identified in the Preamble in accordance with paragraph 5.20 shall identify the body of water. **A3** The location and limits of *excavation for foundations* shall be stated in item descriptions where the limits would otherwise be uncertain. Excavation around pile shafts and for underpinning shall each be so described and classed as *excavation for foundations*. **A4** The Commencing Surface shall be identified in the description of each item for work involving excavation for which the Commencing Surface is not the Original Surface. The Excavated Surface shall be identified in the description of each item for work involving excavation for which the Excavated Surface is not the Final Surface. **A5** Item descriptions for excavation within borrow pits shall so state. **A6** Item descriptions shall identify separately excavation which is expressly required to be carried out by hand. **A7** Item descriptions for *controlled and hazardous materials* shall state the nature of the material.

CLASS E

FIRST DIVISION	SECOND DIVISION	THIRD DIVISION
5 Excavation ancillaries	**1** Trimming of excavated surfaces m^2 **2** Preparation of excavated surfaces m^2 **3** Disposal of excavated material m^3 **4** Double handling of excavated material m^3	**1** Topsoil **2** Material other than topsoil, rock or artificial hard material **3** Rock **4** Stated artificial hard material **5** Controlled and hazardous materials
	5 Dredging to remove silt m^3 **6** Excavation of material below the Final Surface and replacement with stated material m^3 **7** Timber supports left in m^2 **8** Metal supports left in m^2	

MEASUREMENT RULES	DEFINITION RULES	COVERAGE RULES	ADDITIONAL DESCRIPTION RULES
M10 *Trimming of excavated surfaces* shall be measured to surfaces which are to receive no Permanent Works whether trimming is expressly required or not. **M11** *Preparation of excavated surfaces* shall be measured to surfaces which are to receive Permanent Works whether preparation is expressly required or not except surfaces which are to receive filling or landscaping and surfaces for which formwork is measured. **M12** The volume of *disposal of excavated material* measured shall be the difference between the total net volume of excavation and the net volume of excavated material used for filling. **M13** *Double handling of excavated material* shall be measured only to the extent that it is expressly required. The volume measured for double handling shall be that of the void formed in the temporary stockpile from which the material is removed. **M14** *Dredging to remove silt* shall be measured only to the extent that it is expressly required that silt which accumulates after the Final Surface has been reached shall be removed. **M15** The area measured for *timber* or *metal supports left in* shall be the area of supported surfaces for which the supports are expressly required to be left in.	**D5** *Disposal of excavated material* shall be deemed to be disposal off the Site unless otherwise stated in item descriptions. **D6** *Trimming, preparation, disposal* and *double handling* shall be deemed to be carried out upon material other than topsoil, rock or artificial hard material unless otherwise stated in item descriptions.		**A8** Item descriptions for *excavation ancillaries* in connection with excavation by dredging shall be so described. **A9** Item descriptions for *trimming of excavated surfaces* and *preparation of excavated surfaces* shall identify surfaces which are: (*a*) inclined at an angle of 10°–45° to the horizontal (*b*) inclined at an angle of 45°–90° to the horizontal (*c*) vertical. **A10** Where material is for disposal on the Site the location of the disposal areas shall be stated in item descriptions for *disposal of excavated material*. **A11** Item descriptions for *disposal of controlled and hazardous materials* shall state the nature of the material.

CLASS E

FIRST DIVISION	SECOND DIVISION		THIRD DIVISION	
6 Filling	**1** To structures **2** Embankments **3** General **4** To stated depth or thickness	m^3 m^3 m^3 m^2	**1** Excavated topsoil **2** Imported topsoil **3** Non-selected excavated material other than topsoil or rock **4** Selected excavated material other than topsoil or rock **5** Imported natural material other than topsoil or rock **6** Excavated rock **7** Imported rock **8** Imported artificial material	
	1 High energy impact compaction general fill		**1** Selected excavated material other than topsoil or rock **2** Imported natural material other than topsoil or rock **3** Existing ground	m^3 m^3 m^2
7 Filling ancillaries	**1** Trimming of filled surfaces **2** Preparation of filled surfaces	m^2 m^2	**1** Topsoil **2** Material other than topsoil, rock or artificial hard material **3** Rock **4** Stated artificial hard material	
	3 Geotextiles	m^2		
8 Landscaping	**1** Turfing **2** Hydraulic mulch grass seeding **3** Other grass seeding **4** Plants, stated species and size **5** Shrubs, stated species and size **6** Trees, stated species and size	m^2 m^2 m^2 nr nr nr		
	7 Hedges, stated species, size and spacing	m	**1** Single row **2** Double row	

MEASUREMENT RULES	DEFINITION RULES	COVERAGE RULES	ADDITIONAL DESCRIPTION RULES
M16 *Filling* of excavations around completed structures shall be measured only to the extent that the volume filled is also measured as excavation in accordance with rule M6. **M17** Where *filling* to form temporary roads is subsequently approved by the contract administrator for incorporation into permanent filling the volume placed shall not be deducted from the measurement of filling. **M18** Additional *filling* necessitated by settlement of or penetration into underlying material shall be measured only to the extent that its depth exceeds 75 mm. **M19** The volume of *imported filling* material measured shall be the difference between the net volume of filling and the net volume of excavated material derived from work within classes E and T used for filling. **M20** Where *rock filling* is deposited into soft areas the volume shall be measured in the transport vehicles at the place of deposition. **M21** Where *filling* is to be deposited below water, and the quantity cannot be measured satisfactorily by any other means, its volume shall be measured in the transport vehicles at the place of deposition.	**D7** Filling material shall be deemed to be *non-selected excavated material other than topsoil or rock,* unless otherwise stated in item descriptions. **D8** Filling material shall be classed as *excavated rock* only where the use of rock as filling at stated locations is expressly required. **D9** Filling shall be classed as *to stated depth or thickness* where material is provided of uniform total compacted depth or thickness such as in drainage blankets, topsoiling, pitching and beaching. Bulk filling shall not be classed as *to stated depth or thickness* notwithstanding that it may be compacted in separate layers of material of stated thickness.	**C4** Items for *filling* shall be deemed to include compaction.	**A12** The materials shall be identified in item descriptions for filling with *imported material*. **A13** Where different compaction requirements are specified for the same filling material they shall be stated in item descriptions for *filling*. **A14** Where the rate of deposition of filling material is limited the limitation shall be stated in item descriptions for *filling*. **A15** The materials shall be identified in item descriptions for *filling to stated depth or thickness*. **A16** Item descriptions for *filling to stated depth or thickness* shall identify work upon surfaces which are: (*a*) inclined at an angle of 10°–45° to the horizontal (*b*) inclined at an angle of 45°–90° to the horizontal (*c*) vertical. **A17** Item descriptions for *high energy impact compaction* shall state the type of compaction.
M22 High energy impact compaction general fill existing ground shall be measured to the surface area to be treated			
M23 *Trimming of filled surfaces* shall be measured to surfaces which are to receive no Permanent Works whether trimming is expressly required or not. **M24** *Preparation of filled surfaces* shall be measured to surfaces which are to receive Permanent Works whether preparation is expressly required or not except surfaces which are to receive filling or landscaping and surfaces for which formwork is measured.	**D10** *Trimming* and *preparation* shall be deemed to be carried out upon material other than topsoil, rock or artificial hard material unless otherwise stated in item descriptions.		**A18** Item descriptions for *filing ancillaries* shall identify work upon surfaces which are: (*a*) inclined at an angle of 10°–45° to the horizontal (*b*) inclined at an angle of 45°–90° to the horizontal (*c*) vertical. **A19** The type and grade of material shall be stated in item descriptions for *geotextiles*.
M25 The area of additional *geotextiles* in laps shall not be measured.			
M26 The lengths of *hedges* measured shall be their developed lengths along centre lines.		**C5** Items for *landscaping* shall be deemed to include fertilizing, trimming and preparation of surfaces	**A20** Where *turfing* is pegged or wired item descriptions shall so state. **A21** Item descriptions for *turfing* and *grass seeding* shall identify work upon surfaces which are inclined at an angle exceeding 10° to the horizontal.

CLASS F: IN SITU CONCRETE

Excludes: In situ concrete for:
capping of boreholes (included in class B)
diaphragm walls (included in class C)
excavation ancillaries (included in class E)
granolithic and other applied finishes (included in class G)
drainage and pipework (included in classes K and L)
piles (included in classes P and Q)
roads, pavings and kerbs (included in class R)
tunnel and shaft linings (included in class T)
foundations for fences and gates (included in class X)

FIRST DIVISION		SECOND DIVISION	THIRD DIVISION
Provision of concrete **1** Designed concrete	m^3	**1** Strength C8/10 **2** C12/15 **3** C16/20 **4** C20/25 **5** C25/30 **6** C28/35 **7** C30/37 **8** C32/40	**1** Maximum aggregate size 10 mm **2** 20 mm **3** 40 mm
Provision of concrete **2** Designated concrete	m^3	**1** Strength RC20/25 **2** RC25/30 **3** RC28/35 **4** RC30/37 **5** RC32/40 **6** RC35/40 **7** RC40/50	
Provision of concrete **3** Standardised prescribed concrete	m^3	**1** ST1 **2** ST2 **3** ST3 **4** ST4 **5** ST5	
Provision of concrete **4** Prescribed mix	m^3		
Provision of concrete **5** Proprietary concrete	m^3		

MEASUREMENT RULES	DEFINITION RULES	COVERAGE RULES	ADDITIONAL DESCRIPTION RULES
M1 The volume of concrete measured shall include that occupied by (*a*) reinforcement and other metal sections (*b*) prestressing components (*c*) cast-in components each not exceeding 0.1 m^3 in volume (*d*) rebates, grooves, throats, fillets, chamfers or internal splays each not exceeding 0.01 m^2 in cross-sectional area (*e*) pockets and holes which are defined as large or small voids in accordance with rule D3 of class G (*f*) joints or joint components between adjacent volumes of in situ concrete. **M2** The volume of concrete measured shall exclude that of nibs or external splays each not exceeding 0.01 m^2 in cross-sectional area.			
	D1 Items for the provision of concrete shall be classified in accordance with BS 8500 and (November 2006) BS EN 206-1. **D2** Concrete shall be classed as *Designed concrete* where the designer is responsible for ensuring the requirements for the *concrete* are in accordance with the Standards.		**A1** The specification of concrete in accordance with BS 8500 shall be stated in the item description for *Provision of concrete* unless the concrete reference is stated for which the specification is given elsewhere in the Contract. **A2** Where BS 8500 is not used the specification or item description shall state the details of the concrete to be provided.
	D3 Concrete shall be classed as *Designated concrete* where the design of the concrete is generated by the concrete producer.		**A3** The specified design exposure conditions shall be stated in the item description for *Provision of concrete* unless the concrete reference is stated for which the specification is given elsewhere in the Contract.
	D4 Concrete shall be classed as *Standard prescribed concrete* where it is limited to simple structural non reinforced applications		**A4** Any specified requirements for admixtures shall be stated in the item description for *Provision of concrete* unless the concrete reference is stated for which the specification is given elsewhere in the Contract.
	D5 Concrete shall be classed as *Prescribed concrete* where the designer provides full details of the concrete such that it will produce concrete that will achieve the required performance.		**A5** Any specified requirements for 'Consistence categories' shall be stated in the item description for *Provision of concrete* unless the concrete reference is stated for which the specification is given elsewhere in the Contract. **A6** Any specified requirements for 'Maximum water to cement ratios' shall be stated in the item description for *Provision of concrete* unless the concrete reference is stated for which the specification is given elsewhere in the Contract. **A7** Any specified requirements for 'Minimum cement content kg/m^3' shall be stated in the item description for *Provision of concrete* unless the concrete reference is stated for which the specification is given elsewhere in the Contract.
	D6 Concrete shall be classed as *Proprietary concrete* where the application is to achieve a particular performance outside normal performance criteria.		

CLASS F

FIRST DIVISION		SECOND DIVISION	THIRD DIVISION
Placing of concrete **6** Mass **7** Reinforced **8** Prestressed	 m^3 m^3 m^3	**1** Blinding **2** Bases, footings, pile caps and ground slabs **3** Suspended slabs **4** Walls	**1** Thickness: not exceeding 150 mm **2** 150–300 mm **3** 300–500 mm **4** exceeding 500 mm
		5 Columns and piers **6** Beams **7** Casing to metal sections	**1** Cross-sectional area: not exceeding $0.03\ m^2$ **2** $0.03–0.1\ m^2$ **3** $0.1–0.25\ m^2$ **4** $0.25–1\ m^2$ **5** exceeding $1\ m^2$ **6** Special beam sections
		8 Other concrete forms	

NOTE

In accordance with paragraph 5.10 the location of concrete members in the Works may be stated in item descriptions for *placing of concrete* where special characteristics may affect the method and rate of placing concrete.

MEASUREMENT RULES	DEFINITION RULES	COVERAGE RULES	ADDITIONAL DESCRIPTION RULES
M3 *Columns and piers* integral with a wall shall be measured as part of the wall, except where expressly required to be cast separately. **M4** *Beams* integral with a slab shall be measured as part of the slab, except where expressly required to be cast separately.	**D7** Prestressed concrete which is also reinforced shall be classed as *prestressed concrete*. **D8** The thickness used for classification of *blinding* shall be the minimum thickness. **D9** The thickness used for classification of *ground slabs, suspended slabs* and *walls* shall exclude the additional thickness of integral beams, columns, piers and other projections. **D10** Concrete in *suspended slabs* and *walls* less than 1 m wide or long shall be classed as concrete in *beams* and *columns* respectively. **D11** Beams shall be classed as *special beam sections* where their cross-section profiles are rectangular or approximately rectangular over less than 4/5 of their length or where they are of box or other composite section. **D12** Sprayed concrete shall be designated as other concrete forms. **D13** Reinforcing materials added to the mix for sprayed concrete shall not be classed as reinforcement.		**A8** Item descriptions for *placing of concrete* which is expressly required to be placed against an excavated surface (other than blinding) shall so state. **A9** The cross-sectional dimensions of *special beam sections* shall be stated in item descriptions, except where a beam type or mark number is stated for which dimensions are given on the drawings. **A10** Item descriptions for components classed as *other concrete forms* shall identify the work and include one of the following: (*a*) the principal dimensions (*b*) a type or mark number for which principal dimensions are given on the drawings (*c*) a statement locating the work for which principal dimensions are given on the drawings. **A11** Item descriptions for *sprayed concrete support* shall state the specification of the concrete and whether it is reinforced and the minimum thickness.

CLASS G: CONCRETE ANCILLARIES

Includes: **Formwork for in situ concrete**
Reinforcement for in situ concrete
Joints in in situ concrete
Post-tensioned prestressing
Accessories for in situ concrete

FIRST DIVISION	SECOND DIVISION	THIRD DIVISION
1 Formwork: rough finish **2** Fair finish **3** Other stated finish **4** Stated surface features	**1** Plane horizontal **2** Plane sloping **3** Plane battered **4** Plane vertical **5** Curved to one radius in one plane	**1** Width: not exceeding 0.1 m m **2** 0.1–0.2 m m **3** 0.2–0.4 m m^2 **4** 0.4–1.22 m m^2 **5** exceeding 1.22 m m^2
	6 Other curved m^2	
	7 For voids nr	**1** Small void depth: not exceeding 0.5 m **2** 0.5–1 m **3** 1–2 m **4** stated exceeding 2 m **5** Large void depth: not exceeding 0.5 m **6** 0.5–1 m **7** 1–2 m **8** stated exceeding 2 m
	8 For concrete components of constant cross-section m	**1** Beams **2** Columns **3** Walls **4** Other members **5** Projections **6** Intrusions
5 Reinforcement	**1** Plain round steel bars t **2** Deformed high yield steel bars t **3** Stainless steel bars of stated quality t **4** Reinforcing bars of other stated material t	**1** Nominal size: 6 mm **2** 8 mm **3** 10 mm **4** 12 mm **5** 16 mm **6** 20 mm **7** 25 mm **8** 32 mm or greater
	5 Special joints nr	
	6 Steel fabric m^2 **7** Fabric of other stated material m^2	**1** Nominal mass: not exceeding 2 kg/m^2 **2** 2–3 kg/m^2 **3** 3–4 kg/m^2 **4** 4–5 kg/m^2 **5** 5–6 kg/m^2 **6** 6–7 kg/m^2 **7** 7–8 kg/m^2 **8** stated exceeding 8 kg/m^2

Excludes: **Reinforcement in diaphragm walls (included in class C)**
Pre-tensioned prestressing (included in class H)
Formwork and reinforcement in precast concrete (included in class H)
Formwork and reinforcement ancillary to drainage and pipework (included in classes K and L)
Formwork and reinforcement in piles (included in classes P and Q)
Formwork and reinforcement for concrete roads and pavings (included in class R)
Formwork for tunnel and shaft linings (included in class T)
Formwork for foundations for fences and gates (included in class X)

MEASUREMENT RULES

M1 *Formwork* shall be measured for surfaces of in situ concrete which require temporary support during casting except where otherwise stated in CESMM3.

M2 *Formwork* shall not be measured for the following:
(*a*) edges of blinding concrete not exceeding 0.2 m wide
(*b*) joints and associated rebates and grooves
(*c*) temporary surfaces formed at the discretion of the Contractor
(*d*) surfaces of concrete which are expressly required to be cast against an excavated surface
(*e*) surfaces of concrete which are cast against excavated surfaces inclined at an angle less than 45° to the horizontal.

M3 *Formwork* to upper surfaces of concrete shall be measured to surfaces inclined at an angle exceeding 15° to the horizontal and to other upper surfaces for which formwork is expressly required.

M4 *Formwork* for the surfaces of voids larger than those classed as large voids in accordance with rule D3 shall be measured as set out in this class for formwork generally.

M5 *Formwork* for the surfaces of projections and intrusions exceeding 0.01 m^2 in cross-sectional area shall be measured as set out in this class for formwork generally.

M6 The area of *formwork* measured shall include the area of formwork obscured by:
(*a*) forms for *large* and *small voids*
(*b*) forms for *projections* and *intrusions*
(*c*) inserts

M7 The mass of steel *reinforcement* shall be taken as 0.785 kg/m per 100 mm^2 of cross-section (7.85 t/m^3). The mass of other reinforcing materials shall be taken as stated in the Contract.

M8 The mass of *reinforcement* measured shall include the mass of steel supports to top reinforcement.

M9 The area of additional *fabric* in laps shall not be measured.

DEFINITION RULES

D1 *Plane formwork* shall be classified according to its angle of inclination as follows:

Class	Angle of inclination to the vertical
Horizontal	85°–90°
Sloping	10°–85°
Battered	0°–10°
Vertical	0°

D2 *Formwork* shall be deemed to be for plane areas and to exceed 1.22 m wide, unless otherwise stated.

D3 The classification of *large and small voids* shall be as follows:

Class	Maximum cross-section	
	Circular voids (diameter)	Other voids (area)
Large	0.35–0.7 m	0.1–0.5 m^2
Small	Not exceeding 0.35 m	Not exceeding 0.1 m^2

The depths of voids shall be measured perpendicularly to the adjacent surface of concrete.

D4 Nibs and external splays not exceeding 0.01 m^2 in cross-sectional area shall be classed as *projections*.

D5 Rebates, grooves, internal splays, throats, fillets and chamfers not exceeding 0.01 m^2 in cross-sectional area shall be classed as *intrusions*.

D6 Welded, swaged or screwed sleeve joints in reinforcing bars shall be classed as *special joints*.

COVERAGE RULES

C1 Items for *reinforcement* shall be deemed to include supporting reinforcement other than steel supports to top reinforcement.

ADDITIONAL DESCRIPTION RULES

A1 Formwork left in shall be so described in item descriptions for *formwork*.

A2 Item descriptions for *formwork* which is to upper surfaces shall so state, except where the surfaces are inclined at an angle not exceeding 10° to the vertical.

A3 Item descriptions for *formwork* shall state where the formwork is to blinding concrete.

A4 Radii of *curved formwork* shall be stated in item descriptions as follows:
(*a*) to one radius in one plane (cylindrical), radius stated
(*b*) to one radius in two planes (spherical), radius stated
(*c*) varying radius (conical), maximum and minimum radii stated.

A5 Item descriptions for formwork *for concrete components of constant cross-section,* other than *projections* and *intrusions,* shall state the principal cross-sectional dimensions of the component and its mark number, location or other unique identifying feature.

A6 *Formwork* for curved *concrete components of constant cross-section* shall be so described stating the radii.

A7 Item descriptions for bar *reinforcement* shall state the lengths of bars to the next higher multiple of 3 m where they exceed 12 m before bending.

A8 Item descriptions for *special joints* shall state the type of joint and type and size of reinforcing bar.

A9 Item descriptions for *steel fabric* shall state the fabric reference size and mass or mass per square metre.

A10 Item descriptions for *fabric of other stated material* shall state the sizes and nominal mass per square metre.

CLASS G

FIRST DIVISION	SECOND DIVISION	THIRD DIVISION
6 Joints	**1** Open surface plain m^2 **2** Open surface with filler m^2 **3** Formed surface plain m^2 **4** Formed surface with filler m^2	**1** Average width: not exceeding 0.5 m **2** 0.5–1 m **3** stated exceeding 1 m
	5 Plastics or rubber waterstops m **6** Metal waterstops m	**1** Average width: not exceeding 150 mm **2** 150–200 mm **3** 200–300 mm **4** stated exceeding 300 mm
	7 Sealed rebates or grooves m	
	8 Dowels nr	**1** Plain or greased **2** Sleeved or capped
7 Post-tensioned prestressing nr	**1** Horizontal internal tendons in in situ concrete **2** Inclined or vertical internal tendons in in situ concrete **3** Horizontal internal tendons in precast concrete **4** Inclined or vertical internal tendons in precast concrete	**1** Length: not exceeding 5 m **2** 5–7 m **3** 7–10 m **4** 10–15 m **5** 15–20 m **6** 20–25 m **7** 25–30 m **8** stated exceeding 30 m
	5 External jacking operations	
8 Concrete accessories	**1** Finishing of top surfaces m^2	**1** Wood float **2** Steel trowel **3** Other stated surface treatment **4** Granolithic finish **5** Other stated applied finish
	2 Finishing of formed surfaces m^2	**1** Aggregate exposure using retarder **2** Bush hammering **3** Other stated surface treatment carried out after striking formwork
	3 Inserts	**1** Linear inserts m **2** Other inserts nr
	4 Grouting under plates nr	**1** Area: not exceeding 0.1 m^2 **2** 0.1–0.5 m^2 **3** 0.5–1 m^2 **4** stated exceeding 1 m^2

NOTE

Similar *inserts* which vary in size may be added together and classified by size within ranges.

CLASS G

MEASUREMENT RULES	DEFINITION RULES	COVERAGE RULES	ADDITIONAL DESCRIPTION RULES
M10 *Joints* shall be measured only where they are at locations where joints are expressly required. **M11** The widths of *joints* shall be measured between the outer surfaces of concrete with no deduction or addition for widths or depths occupied by rebates, grooves, fillets or waterstops. The lengths of *waterstops* shall be measured along their centre lines.	**D7** *Joints* for which temporary support of the whole surface area of concrete is required during casting shall be classed as *formed surface joints*. Other joints shall be classed as *open surface joints*.	**C2** Items for *open surface* and *formed surface joints* shall be deemed to include intermediate surface treatment where expressly required. **C3** Items for *joints* shall be deemed to include formwork. **C4** Items for *waterstops* shall be deemed to include cutting and joining of waterstops and provision of special fittings at angles and junctions.	**A11** The dimensions and nature of components shall be stated in item descriptions for *joints*.
M12 *Prestressing* shall be measured by the number of *tendons* where tendons are used and by the number of *external jacking operations* where stress is induced by jacking only.	**D8** Profiled *tendons* in horizontal components shall be classed as *horizontal tendons*. **D9** The lengths of *tendons* used for classification shall be their developed lengths between the outer faces of anchorages.	**C5** Items for *prestressing* shall be deemed to include ducts, grouting and other components and tasks ancillary to prestressing.	**A12** Item descriptions for *prestressing* shall identify the concrete component to be stressed and state the composition of the tendon and particulars of the anchorage.
M13 Surface treatments shall not be measured to surfaces formed at the Contractor's discretion. **M14** The areas of tops of walls and other surfaces which are not given separate finishing treatment shall not be measured as *finishing of top surfaces*. **M15** No deduction from the areas measured for *finishing* shall be made for holes and openings in the finished surfaces each not exceeding 0.5 m^2.		**C6** Items for *granolithic* and *other stated applied finish* shall be deemed to include materials, surface treatment, joints and formwork.	**A13** The materials, thicknesses and surface treatments of *granolithic* and *other stated applied finish* shall be stated in item descriptions.
M16 Where *inserts* are expressly required to be grouted into preformed openings the formwork shall be measured.	**D10** Components cast or grouted into in situ concrete except reinforcement, prestressing and jointing materials shall be classed as *inserts*.	**C7** Items for *inserts* shall be deemed to include their supply unless otherwise stated.	**A14** Item descriptions for *inserts* shall identify the components to be cast or grouted in and state their principal dimensions. **A15** Item descriptions for *inserts* shall identify: (*a*) those which project from one surface of the concrete (*b*) those which project from two surfaces of the concrete (*c*) those which are totally within the concrete volume. **A16** Where *inserts* are expressly required to be grouted into preformed openings in concrete item descriptions shall so state. Materials for grouting and sizes of openings shall be stated.
			A17 Materials shall be stated in item descriptions for *grouting under plates*.

CLASS H: PRECAST CONCRETE

Includes: **Manufacture, erection, joining and fixing of precast concrete units**
Excludes: **Post-tensioned prestressing (included in class G)**
Precast concrete pipework (included in classes I and J)
Precast concrete manholes, catchpits and gullies (included in class K)
Precast concrete piles (included in classes P and Q)
Precast concrete paving, kerbs and traffic sign supports (included in class R)
Precast concrete tunnel linings (included in class T)
Precast concrete blockwork (included in class U)
Precast concrete fencing (included in class X)

FIRST DIVISION		SECOND DIVISION	THIRD DIVISION
1 Beams **2** Prestressed pre-tensioned beams **3** Prestressed post-tensioned beams **4** Columns	nr nr nr nr	**1** Length: not exceeding 5 m **2** 5–7 m **3** 7–10 m **4** 10–15 m **5** 15–20 m **6** 20–30 m **7** exceeding 30 m	**1** Mass: not exceeding 250 kg **2** 250–500 kg **3** 500 kg–1 t **4** 1–2 t **5** 2–5 t **6** 5–10 t **7** 10–20 t **8** stated exceeding 20 t
5 Slabs	nr	**1** Area: not exceeding 1 m^2 **2** 1–4 m^2 **3** 4–15 m^2 **4** 15–50 m^2 **5** exceeding 50 m^2	
6 Segmental units	nr		
7 Units for subways, culverts and ducts	m		
8 Copings, sills and weir blocks	m	**1** Cross-sectional area: not exceeding 0.1 m^2 **2** 0.1–0.5 m^2 **3** 0.5–1 m^2 **4** exceeding 1 m^2	

MEASUREMENT RULES	DEFINITION RULES	COVERAGE RULES	ADDITIONAL DESCRIPTION RULES
	D1 The mass used for classification in the third division shall be the mass of each unit. **D2** Concrete components which are cast other than in their final position shall generally be classed as *precast concrete* units. **D3** Where site precasting of units is adopted for reasons other than to obtain multiple use of formwork and the nature of the work is characteristic of in situ concrete, but involves the movement of the cast units into their final positions, the units shall be classed as *in situ concrete* and items given in class A for the Temporary Works associated with the movement of the units.	**C1** Items for precast concrete shall be deemed to include reinforcement, formwork, joints and finishes.	**A1** The position in the Works and specification of the concrete to be used in each type of precast unit shall be stated in item descriptions. **A2** Item descriptions shall state the mark or type number of each precast concrete unit. Units with different dimensions shall be given different mark or type numbers. **A3** Particulars of tendons and prestressing shall be stated in item descriptions for prestressed pre-tensioned units. **A4** The cross-section type and principal dimensions shall be stated in item descriptions for *beams, columns, segmental units, units for subways, culverts, ducts, copings, sills and weir blocks*. **A5** The average thickness shall be stated in item descriptions for *slabs*. **A6** The mass per metre shall be stated in item descriptions for *units for subways, culverts, ducts, copings, sills and weir blocks*.
M1 The length measured for *units for subways, culverts and ducts and for copings, sills and weir blocks* shall be the total length of identical units.			

CLASS I: PIPEWORK – PIPES

Includes: **Provision, laying and jointing of pipes**
Excavating and backfilling pipe trenches
Excludes: **Work included in classes J, K, L and Y**
Piped building services (included in class Z)

FIRST DIVISION		SECOND DIVISION	THIRD DIVISION
1 Clay pipes	m	**1** Nominal bore: not exceeding 200 mm	**1** Not in trenches
2 Concrete pipes	m	**2** 200–300 mm	**2** In trenches, depth: not exceeding 1.5 m
3 Iron pipes	m	**3** 300–600 mm	**3** 1.5–2 m
4 Steel pipes	m	**4** 600–900 mm	**4** 2–2.5 m
5 Polyvinyl chloride pipes	m	**5** 900–1200 mm	**5** 2.5–3 m
6 Glass reinforced plastic pipes	m	**6** 1200–1500 mm	**6** 3–3.5 m
7 High density polyethylene pipes	m	**7** 1500–1800 mm	**7** 3.5–4 m
8 Medium density polyethylene pipes	m	**8** exceeding 1800 mm	**8** exceeding 4 m

MEASUREMENT RULES	DEFINITION RULES	COVERAGE RULES	ADDITIONAL DESCRIPTION RULES
M1 The Commencing Surface adopted in the preparation of the Bill of Quantities shall be adopted for the measurement of the completed work. **M2** Backfilling of trenches shall not be measured except as set out in class K for filling of French and rubble drains and in class L for backfilling with material other than that excavated from the trenches. **M3** Lengths of pipes shall be measured along their centre lines. Lengths of pipes *in trenches* shall include lengths occupied by fittings and valves and exclude lengths occupied by pipes and fittings comprising backdrops to manholes. Lengths of pipes *not in trenches* shall exclude lengths occupied by fittings and valves. Where more than one pipe is expressly required to be laid in one trench, the length measured shall be the length of the pipe, not including the length occupied by manholes and other chambers. **M4** Additional items shall be given in classes K and L for work in connection with pipes *not in trenches* other than the provision, laying and jointing of pipes. **M5** Lengths of pipes entering manholes and other chambers shall be measured to the inside surfaces of the chambers except that pipes and fittings comprising backdrops to manholes shall be included in items for manholes measured in class K.	**D1** Pipes *not in trenches* shall include pipes suspended or supported above the ground or other surface, pipes in headings, tunnels or shafts, pipes installed by thrust boring and pipe jacking and pipes laid within volumes measured separately for excavation. **D2** Pipes *not in trenches* shall be classed as such only where pipes are expressly required not to be laid in trenches. **D3** Depths used for classification of pipes *in trenches* shall be measured from the Commencing Surface to the inverts of the pipes.		**A1** The location or type of pipework in each item or group of items shall be stated in item descriptions so that the pipe runs included can be identified by reference to the Drawings. **A2** The materials, joint types, nominal bores and lining requirements of pipes shall be stated in item descriptions. **A3** Item descriptions for pipes *not in trenches* shall distinguish between the different categories of pipes listed in rule D1. **A4** The Commencing Surface shall be identified in the description of each item for work involving excavation for which the Commencing Surface is not also the Original Surface. **A5** Where more than one pipe is expressly required to be laid in one trench the item descriptions for each pipe shall so state and also identify the pipe run. Where pipes are laid in French or rubble drains item descriptions shall so state. **A6** Trench depths exceeding 4 m shall be stated in item descriptions to the next higher multiple of 0.5 m. **A7** Item descriptions shall identify separately excavation which is expressly required to be carried out by hand.

CLASS J: PIPEWORK – FITTINGS AND VALVES

Includes: **Fittings and valves for pipework**
Excludes: **Work included in classes I, K, L and Y**
Piped building services (included in class Z)

FIRST DIVISION		SECOND DIVISION	THIRD DIVISION
1 Clay pipe fittings	nr	**1** Bends	**1** Nominal bore: not exceeding 200 mm
2 Concrete pipe fittings	nr	**2** Junctions and branches	**2** 200–300 mm
3 Iron or steel pipe fittings	nr	**3** Tapers	**3** 300–600 mm
4 Polyvinyl chloride pipe fittings	nr	**4** Double collars	**4** 600–900 mm
5 Glass reinforced plastic pipe fittings	nr	**5** Adaptors	**5** 900–1200 mm
6 High density polyethylene pipe fittings	nr	**6** Glands	**6** 1200–1500 mm
7 Medium density polyethylene pipe fittings	nr	**7** Bellmouths	**7** 1500–1800 mm
		8 Straight specials	**8** exceeding 1800 mm
8 Valves and penstocks	nr	**1** Gate valves: hand operated	
		2 power operated	
		3 Non-return valves	
		4 Butterfly valves: hand operated	
		5 power operated	
		6 Air valves	
		7 Pressure reducing valves	
		8 Penstocks	

MEASUREMENT RULES	DEFINITION RULES	COVERAGE RULES	ADDITIONAL DESCRIPTION RULES
		C1 Items for fittings and valves shall be deemed to include the supply of materials by the Contractor, unless otherwise stated. **C2** Items for fittings and valves shall be deemed to include excavation, preparation of surfaces, disposal of excavated material, upholding sides of excavation, backfilling and removal of dead services except to the extent that such work is included in classes I, K and L.	
M1 *Pipe fittings* comprising backdrops to manholes shall be included in the items for manholes measured in class K. **M2** *Straight specials* shall be measured only where they are expressly required. *Straight specials* shall not be measured where they are necessitated only by the layout of the work. Pipe runs between manholes, chambers and other structures whose lengths are not an exact multiple of a standard pipe length shall be deemed to require *straight specials* only if they are expressly required as rocker pipes.	**D1** *Pipe fittings* on pipes of different nominal bores shall be classified in the third division according to the nominal bore of the largest pipe. **D2** A *straight special* is a length of pipe either cut to length or made to order.	**C3** Items for *straight specials* shall be deemed to include cutting.	**A1** The materials, joint types, nominal bores and lining requirements of *pipe fittings* shall be stated in item. Fittings with puddle flanges shall be so described. **A2** Item descriptions for *pipe fittings* to cast iron or spun iron pipework of nominal bore exceeding 300 mm and to all steel pipework shall state the principal dimensions of each fitting. **A3** Vertical *bends* in pipework of which the nominal bore exceeds 300 mm shall be so described. **A4** *Fittings* to pipework not in trenches shall be so described. **A5** *Fittings* to relined water mains measured in class Y shall be so described.
			A6 The materials, nominal bores and any additional requirements such as joints, draincocks, extension spindles and brackets shall be stated in item descriptions for *valves and penstocks*. **A7** *Valves and penstocks* to relined water mains measured in class Y shall be so described.

CLASS K: PIPEWORK – MANHOLES AND PIPEWORK ANCILLARIES

Includes: **Manholes and other chambers, ducts, culverts, crossings and reinstatement, other ancillaries as listed**
Excludes: **Work included in classes I, J, L and Y**
Ducted building services (included in class Z)

FIRST DIVISION		SECOND DIVISION	THIRD DIVISION
1 Manholes	nr	**1** Brick **2** Brick with backdrop **3** In situ concrete **4** In situ concrete with backdrop **5** Precast concrete **6** Precast concrete with backdrop	**1** Depth: not exceeding 1.5 m **2** 1.5–2 m **3** 2–2.5 m **4** 2.5–3 m **5** 3–3.5 m **6** 3.5–4 m **7** stated exceeding 4 m
2 Other stated chambers	nr	**1** Brick **3** In situ concrete **5** Precast concrete	
3 Gullies	nr	**1** Clay **2** Clay trapped **3** In situ concrete **4** In situ concrete trapped **5** Precast concrete **6** Precast concrete trapped **7** Plastics **8** Plastics trapped	
4 French drains, rubble drains, ditches and trenches		**1** Filling French and rubble drains with graded material m^3 **2** Filling French and rubble drains with rubble m^3	
		3 Trenches for unpiped rubble drains m **4** Rectangular section ditches: unlined m **5** lined m **6** Vee section ditches: unlined m **7** lined m **8** Trenches for pipes or cables not to be laid by the Contractor m	**1** Cross-sectional area: not exceeding 0.25 m^2 **2** 0.25–0.5 m^2 **3** 0.5–0.75 m^2 **4** 0.75–1 m^2 **5** 1–1.5 m^2 **6** 1.5–2 m^2 **7** 2–3 m^2 **8** stated exceeding 3 m^2
5 Ducts and metal culverts	m	**1** Cable ducts: 1 way **2** 2 way **3** 3 way **4** stated number of ways exceeding 3 **5** Sectional corrugated metal culverts, nominal internal diameter: not exceeding 0.5 m **6** 0.5–1 m **7** 1–1.5 m **8** exceeding 1.5 m	**1** Not in trenches **2** In trenches, depth: not exceeding 1.5 m **3** 1.5–2 m **4** 2–2.5 m **5** 2.5–3 m **6** 3–3.5 m **7** 3.5–4 m **8** exceeding 4 m

CLASS K

MEASUREMENT RULES	DEFINITION RULES	COVERAGE RULES	ADDITIONAL DESCRIPTION RULES
	D1 The centre line for multiple pipes, ducts or culverts shall be the line equidistant between the inside faces of the outer pipe walls.	**C1** Items for work in this class shall be deemed to include excavation, preparation of surfaces, disposal of excavated material, upholding sides of excavation backfilling and removal of dead services, except to the extent that such work is included in classes I, J and L. **C2** Items for work in this class shall be deemed to include concrete, reinforcement, formwork, joints and finishes.	
	D2 The *depths* of manholes and other chambers shall be measured from the tops of covers to channel inverts or tops of base slabs, whichever is the lower. **D3** Drawpits shall be classed as *other stated chambers*.	**C3** Item for *manholes, other stated chambers* and *gullies* shall be deemed to include all items of metalwork and pipework, other than valves, which occur within or at the surface of the item. **C4** Items for *manholes* with backdrops shall be deemed to include the pipework and associated fittings comprising the backdrop.	**A1** Type or mark numbers shall be stated in item descriptions for *manholes, other stated chambers* and *gullies* of which details are given elsewhere in the Contract. **A2** Types and loading duties of covers shall be stated in item descriptions for *manholes, other stated chambers* and *gullies*. **A3** Item descriptions for *manholes, other stated chambers* and *gullies* shall identify separately those which are expressly required to be excavated by hand.
M1 Excavation and pipe laying for piped *French and rubble drains* are measured in class I.			**A4** The nature of the filling material shall be stated in item descriptions for *filling French and rubble drains*.
M2 The cross-sectional areas of *lined ditches* shall be measured to the Excavated Surface.			**A5** Materials and dimensions of linings to ditches shall be stated in item descriptions.
M3 The rules in class I for pipes shall also apply to *ducts and metal culverts* in this class except that the lengths measured for ducts and metal culverts not in trenches shall include lengths occupied by fittings.	**D4** Non-circular *metal culverts* shall be classified by their maximum nominal internal cross-sectional dimension. **D5** The rules in class I for pipes shall also apply to *ducts and metal culverts* in this class.	**C5** Items for *ducts and metal culverts* shall be deemed to include cutting and fittings.	**A6** The rules in class I for pipes shall also apply to *ducts and metal culverts* in this class.

CLASS K

FIRST DIVISION		SECOND DIVISION	THIRD DIVISION
6 Crossings	nr	**1** River, stream or canal, width: 1–3 m **2** 3–10 m **3** stated exceeding 10 m **4** Hedge **5** Wall **6** Fence **7** Sewer or drain **8** Other stated underground service	**1** Pipe bore: not exceeding 300 mm **2** 300–900 mm **3** 900–1800 mm **4** stated exceeding 1800 mm
7 Reinstatement	m	**1** Breaking up and temporary reinstatement of roads **2** Breaking up and temporary reinstatement of footpaths **3** Breaking up, temporary and permanent reinstatement of roads **4** Breaking up, temporary and permanent reinstatement of footpaths **5** Reinstatement of land	**1** Pipe bore: not exceeding 300 mm **2** 300–900 mm **3** 900–1800 mm **4** stated exceeding 1800 mm
		6 Strip topsoil from easement and reinstate	
8 Other pipework ancillaries		**1** Reinstatement of field drains m **2** Marker posts nr **3** Timber supports left in excavations m^2 **4** Metal supports left in excavations m^2	
		5 Connections to existing manholes and other chambers nr **6** Connections to existing pipes, ducts and culverts nr	**1** Pipe bore: not exceeding 200 mm **2** 200–300 mm **3** 300–600 mm **4** 600–900 mm **5** 900–1200 mm **6** 1200–1500 mm **7** 1500–1800 mm **8** stated exceeding 1800 mm

NOTE

Manholes and other chambers may be measured in detail as set out in other classes of CESMM3.

CLASS K

MEASUREMENT RULES	DEFINITION RULES	COVERAGE RULES	ADDITIONAL DESCRIPTION RULES
M4 *Crossings* shall be measured for pipes, ducts and metal culverts. **M5** *Crossings of streams* shall be measured only where their width exceeds 1 m.	**D6** *River, stream and canal crossings* shall be classified by their widths measured along pipe, duct or culvert centre lines when the water surface is at the level (or the higher level of fluctuation if applicable) shown on the drawing to which reference is given in the Preamble in accordance with paragraph 5.20. **D7** The dimension used for classification of *bore* in the third division shall be the maximum nominal distance between the inside faces of the outer walls of the pipe, duct or culvert to be installed.	**C6** Items for *crossings* shall be deemed to include reinstatement unless otherwise stated.	**A7** Where linings to *rivers, streams or canals* are to be broken through and reinstated the type of lining shall be stated in item descriptions.
M6 *Reinstatement* shall be measured for pipes, ducts and metal culverts. **M7** Lengths of *reinstatement* shall be measured along centre lines and shall include lengths occupied by manholes and other chambers. **M8** *Strip topsoil from easement and reinstate* shall be measured only where it is expressly required that a width of ground greater than the nominal trench width defined in accordance with rule D1 in class L is to be stripped of topsoil before trench excavation and subsequently reinstated.	**D7** The dimension used for classification of *bore* in the third division shall be the maximum nominal distance between the inside faces of the outer walls of the pipe, duct or culvert to be installed. **D8** Crossings of roads and paths shall be classed as *breaking up and reinstatement* of roads and paths.	**C7** Additional reinstatement shall be deemed to be included in the items for manholes and other chambers. **C8** Removal and reinstatement of kerbs and channels shall be deemed to be included in the items for *breaking up and reinstatement of roads and footpaths.* **C9** Items for *strip topsoil from easement and reinstate* shall be deemed to include storing and protecting topsoil and reinstatement of land.	**A8** Types and depths of surfacing, including base and subbase courses, shall be stated in item descriptions for *breaking up and reinstatement of roads and footpaths.* **A9** Item descriptions for *strip topsoil from easement and reinstate* shall state any limitations on the width to be stripped and reinstated. **A10** Item descriptions for *reinstatement of land* and for *strip topsoil from easement and reinstate* shall distinguish between grassland, gardens, sports fields and cultivated land.
M9 *Other pipework ancillaries* shall be measured for pipes, ducts and metal culverts. **M10** The lengths measured for *reinstatement of field drains* shall be the nominal trench width defined in accordance with rule D1 in class L. **M11** The area measured for *supports left in excavations* shall be the undeveloped area in contact with the surfaces for which the supports are expressly required to be left in.	**D7** The dimension used for classification of *bore* in the third division shall be the maximum nominal distance between the inside faces of the outer walls of the pipe, duct or culvert to be installed.	**C10** Items for *reinstatement of field drains* shall be deemed to include connections to existing field drains.	**A11** Sizes and types of *marker posts* shall be stated in item descriptions. **A12** Item descriptions for *connections to existing manholes and other chambers* and *to existing pipes, ducts and culverts* shall identify the nature of the existing service and the extent of the work to be included.

CLASS L: PIPEWORK – SUPPORTS AND PROTECTION, ANCILLARIES TO LAYING AND EXCAVATION

Includes: **Extras to excavation and backfilling of trenches for pipework, ducts and metal culverts, manholes and other chambers, headings, thrust boring and pipe jacking**
Pipe laying in headings and by thrust boring and pipe jacking
Provision of supports and protection to pipework, ducts and metal culverts

Excludes: **Work included in classes I, J, K and Y**
Insulation to building services (included in class Z)

FIRST DIVISION		SECOND DIVISION	THIRD DIVISION
1 Extras to excavation and backfilling	m^3	**1** In pipe trenches **2** In manholes and other chambers **3** In headings **4** In thrust boring **5** In pipe jacking	**1** Excavation of rock **2** Excavation of mass concrete **3** Excavation of reinforced concrete **4** Excavation of other artificial hard material **5** Backfilling above the Final Surface with concrete **6** Backfilling above the Final Surface with stated material other than concrete **7** Excavation of natural material below the Final Surface and backfilling with concrete **8** Excavation of natural material below the Final Surface and backfilling with stated material other than concrete

MEASUREMENT RULES	DEFINITION RULES	COVERAGE RULES	ADDITIONAL DESCRIPTION RULES
M1 Items shall be given in this class in addition to the items for provision, laying and jointing of *pipes, ducts and culverts* and for the excavation and backfilling of trenches in classes I and K. Items shall be given in this class in addition to the items for manholes and other chambers in classes K and Y. **M2** Work in this class associated with ducts and metal culverts shall be measured as set out for work associated with pipes. The dimension used for classification in the third division (L 2–8 * *) shall be the maximum nominal distance between the inside faces of the outer duct or culvert walls. **M3** Breaking up and reinstatement of roads and pavings shall be included in class K.		**C1** Items for work in this class shall be deemed to include excavation, preparation of surfaces, disposal of excavated material, upholding sides of excavation, backfilling and removal of dead services, except to the extent that such work is included in classes I, J and K or in the items for extras to excavation and backfilling in this class. **C2** Items for work in this class shall be deemed to include concrete, reinforcement, formwork, joints and finishes.	
M4 The volume of extras to excavation and backfilling *in pipe trenches* shall be calculated by multiplying together the average depth and length of the material removed or backfilled and the nominal trench width. **M5** No volume of extras to excavation and backfilling *in manholes and other chambers* shall be measured outside the maximum plan area of the manhole or other chamber. **M6** The volume of extras to excavation and backfilling for pipe laying *in headings, in thrust boring* and *in pipe jacking* shall be measured by multiplying together the internal cross-sectional area of pipe and the length of the material excavated or backfilled. Packing in headings shall not be measured. **M7** *Backfilling above the Final Surface* (L 1 * 5–6) shall be measured only where it is expressly required that the material excavated shall not be used for backfilling. *Excavation below the Final Surface* and backfilling (L 1 * 7–8) shall be measured only where it is expressly required. **M8** An isolated volume of *rock, concrete* or *other artificial hard material* occurring within other material to be excavated shall not be measured separately unless its volume exceeds 0.25 m^3.	**D1** The nominal trench width if not stated in the Contract shall be taken as 500 mm greater than the maximum nominal distance between the inside faces of the outer pipe walls where this distance does not exceed 1 m and as 750 mm greater than this distance where it exceeds 1 m.		**A1** Item descriptions shall identify separately excavation which is expressly required to be carried out by hand.

CLASS L

FIRST DIVISION		SECOND DIVISION	THIRD DIVISION
2 Special pipe laying methods	m	**1** In headings **2** Thrust boring **3** Pipe jacking	**1** Nominal bore: not exceeding 200 mm **2** 200–300 mm **3** 300–600 mm **4** 600–900 mm **5** 900–1200 mm **6** 1200–1500 mm **7** 1500–1800 mm **8** stated exceeding 1800 mm
3 Beds **4** Haunches **5** Surrounds	m m m	**1** Sand **2** Selected excavated granular material **3** Imported granular material **4** Mass concrete **5** Reinforced concrete	
6 Wrapping and lagging	m		
7 Concrete stools and thrust blocks	nr	**1** Volume: not exceeding 0.1 m^3 **2** 0.1–0.2 m^3 **3** 0.2–0.5 m^3 **4** 0.5–1 m^3 **5** 1–2 m^3 **6** 2–4 m^3 **7** 4–6 m^3 **8** stated exceeding 6 m^3	
8 Other isolated pipe supports	nr	**1** Height: not exceeding 1 m **2** 1–1.5 m **3** 1.5–2 m **4** 2–3 m **5** 3–4 m **6** 4–5 m **7** 5–6 m **8** stated exceeding 6 m	

MEASUREMENT RULES	DEFINITION RULES	COVERAGE RULES	ADDITIONAL DESCRIPTION RULES
M9 *Special pipe laying methods* shall be measured only where they are expressly required. **M10** Access pits, shafts and jacking blocks, where expressly stated in the Contract to be executed by the Contractor and of which the nature and extent are expressly stated in the Contract, shall be measured as Specified Requirements in class A.		**C3** Items for *special pipe laying methods* shall be deemed to include crossings, provision and removal of access pits, shafts and jacking blocks unless otherwise stated and other work associated with special pipe laying methods not included in the items for provision, laying and jointing of pipes given in class I.	**A2** Item descriptions for *special pipe laying methods* shall identify the run of pipe or pipes. The type of packing shall be stated in item descriptions for pipes *in headings*.
M11 Lengths of *beds, haunches* and *surrounds* shall be measured along pipe centre lines including lengths occupied by fittings and valves but not including lengths occupied by manholes and other chambers through which they are not continued.	**D2** Items for *haunches* and *surrounds* shall include beds of the same material. **D3** Items for *beds, haunches* and *surrounds* to multiple pipes shall be classified in the third division according to the nominal distance between the inside faces of the outer pipe walls.		**A3** Materials used for *beds, haunches* and *surrounds* and the depths of beds shall be stated in item descriptions. **A4** *Beds, haunches* and *surrounds* to multiple pipes shall be so described stating the number of pipes and the maximum nominal distance between the inside faces of the outer pipe walls.
M12 Lengths of *wrapping and lagging* shall be measured along each pipe centre line including lengths occupied by fittings and valves but not including lengths occupied by manholes and other chambers through which the pipes are not continued.		**C4** Items for *wrapping and lagging of pipes* shall be deemed to include wrapping and lagging of fittings, valves and joints.	**A5** Materials used for *wrapping and lagging* of pipes shall be stated in item descriptions.
	D4 The volumes used for classification of *concrete stools and thrust blocks* shall exclude the volumes occupied by pipes.	**C5** Items for *concrete stools* and *thrust blocks* shall be deemed to include pipe fixings.	**A6** Item descriptions for *concrete stools and thrust blocks* shall state the specification of the concrete and whether it is reinforced.
	D5 The height of *pipe supports* used for classification shall be measured from the ground or other supporting surface to the invert of the highest pipe where pipes are supported from below and of the lowest pipe where pipes are supported from above. **D6** Where two or more pipes are carried by one support, the item for the support shall be classified in the third division by the aggregate bore of the pipes supported.		**A7** Principal dimensions and materials shall be stated in item descriptions for *pipe supports*.

CLASS M: STRUCTURAL METALWORK

Excludes: **Metalwork in concrete (included in classes C, G and H)**
Metalwork in pipework (included in classes I, J, K and L)
Miscellaneous metalwork (included in class N)
Metalwork in piles (included in classes P and Q)
Metalwork in fences (included in class X)

FIRST DIVISION	SECOND DIVISION	THIRD DIVISION
1 Fabrication of main members for bridges t	**1** Rolled sections **2** Plates or flats **3** Built-up box or hollow sections	**1** Straight on plan **2** Curved on plan **3** Straight on plan and cambered **4** Curved on plan and cambered
2 Fabrication of subsidiary members for bridges t	**1** Deck panels	
	2 Bracings **3** External diaphragms	
3 Fabrication of members for frames **4** Fabrication of other members	**1** Columns t **2** Beams t **3** Portal frames t **4** Trestles, towers and built-up columns t **5** Trusses and built-up girders t **6** Bracings, purlins and cladding rails t	**1** Straight on plan **2** Curved on plan **3** Straight on plan and cambered **4** Curved on plan and cambered
	7 Grillages t **8** Anchorages and holding down bolt assemblies nr	
5 Erection of members for bridges **6** Erection of members for frames **7** Erection of other members	**1** Trial erection t **2** Permanent erection t	
	3 Site bolts: black nr **4** HSFG general grade nr **5** HSFG higher grade nr **6** SFG load indicating or load limit types, general grade nr **7** HSFG load indicating or load limit types, higher grade nr	**1** Diameter: not exceeding 16 mm **2** 16–20 mm **3** 20–24 mm **4** 24–30 mm **5** 30–36 mm **6** 36–42 mm **7** stated exceeding 42 mm
8 Off site surface treatment m^2	**1** Blast cleaning **2** Pickling **3** Flame cleaning **4** Wire brushing **5** Metal spraying **6** Galvanizing **7** Painting	

CLASS M

MEASUREMENT RULES	DEFINITION RULES	COVERAGE RULES	ADDITIONAL DESCRIPTION RULES
M1 Items shall be included in miscellaneous metalwork (class N) for metal components not included in this class but associated with metal structures.			
M2 The mass of members, other than *plates or flats*, shall be calculated from the overall lengths of the members with no deductions for splay-cut or mitred ends. **M3** The mass of members measured shall be that of plates, rolled sections, shear connectors, stiffeners, cleats, packs, splice plates and other fittings. **M4** No allowance shall be made in the measurements for rolling margin and other permissible deviations. The mass of weld fillets, bolts, nuts, washers, rivets and protective coatings shall not be measured. **M5** No deductions shall be made for the mass of metal removed to form notches and holes each not exceeding 0.1 m^2 in area measured in plane. **M6** The mass of steel shall be taken for measurement as 785 kg/m^2 per 100 mm thickness (7.85 t/m^3). The masses of other metals shall be taken as stated in the Specification or, where not so stated, as stated in the supplier's catalogue. **M7** *Anchorages and holding down bolt assemblies* shall be measured by the number of complete assemblies.		**C1** Items for *fabrication* of metalwork shall be deemed to include delivery of fabricated metalwork to the Site.	**A1** The materials and grades of materials used shall be stated in item descriptions for *fabrication of members*. **A2** Item descriptions for *fabrication of members* shall identify tapered or castellated members. **A3** Item descriptions for *fabrication of members* other than for portal frames shall identify cranked members. **A4** Details of the members comprising boom and infill construction shall be stated in item descriptions for *trestles, towers and built-up columns* and *trusses and built-up girders*. **A5** Item descriptions for *anchorages and holding down bolt assemblies* shall state particulars of the type of anchorage or assembly.
		C2 Items for *erection of members* shall be deemed to include work carried out after delivery of fabricated metalwork to the Site. **C3** Items for *site bolts* shall be deemed to include supply and delivery to the Site.	**A6** Item descriptions for *erection* shall separately identify and locate separate bridges and structural frames and, where appropriate, parts of bridges or frames. **A7** Where fixing clips and resilient pads are used to secure overhead crane rails, item descriptions shall so state.
M8 Surface treatment carried out on the Site shall be classed as painting (class V).			**A8** Materials and number of applications shall be stated in item descriptions for *metal spraying, galvanizing* and *painting*.

CLASS N: MISCELLANEOUS METALWORK

Excludes: **Metal reinforcement for concrete (included in classes C, G, H and R)**
Metal inserts in concrete (included in classes G and H)
Pipework (included in classes I, J, K and L)
Structural metalwork (included in class M)
Fittings and fastenings to timber (included in class O)
Piles (included in classes P and Q)
Traffic signs (included in class R)
Rail track and accessories (included in class S)
Fences (included in class X)

FIRST DIVISION	SECOND DIVISION		THIRD DIVISION
1	**1** Stairways and landings **2** Walkways and platforms	t t	
	3 Ladders **4** Handrails **5** Bridge parapets	m m m	
	6 Miscellaneous framing	m	**1** Angle section **2** Channel section **3** I section **4** Tubular section
	7 Plate flooring **8** Open grid flooring	m^2 m^2	
2	**1** Cladding **2** Welded mesh panelling **3** Duct covers	m^2 m^2 m^2	
	4 Tie rods	nr	
	5 Walings	m	
	6 Bridge bearings	nr	**1** Roller **2** Slide **3** Rocker **4** Cylindrical **5** Spherical **6** Elastomeric **7** Rubber pot **8** Proprietary
	7 Uncovered tanks **8** Covered tanks	nr nr	**1** Volume: not exceeding 1 m^3 **2** 1–3 m^3 **3** 3–10 m^3 **4** 10–30 m^3 **5** 30–100 m^3 **6** 100–300 m^3 **7** 300–1000 m^3 **8** stated exceeding 1000 m^3

NOTE

As an alternative to the additional description given as required by rule A1, item descriptions may identify assemblies by mark number in accordance with paragraph 5.12.

MEASUREMENT RULES	DEFINITION RULES	COVERAGE RULES	ADDITIONAL DESCRIPTION RULES
M1 Painting carried out on the Site shall be classed as painting (class V). **M2** Masses calculated for miscellaneous metalwork assemblies shall include the mass of all metal components and attached pieces. **M3** No deduction from the masses or areas measured for miscellaneous metalwork shall be made for openings and holes each not exceeding 0.5 m^2 in area.		**C1** Items for miscellaneous metalwork shall be deemed to include fixing to other work, supply of fixing components and drilling or cutting of other work.	**A1** Item descriptions shall state the specification and thicknesses of metal, surface treatments carried out prior to delivery to site and principal dimensions of miscellaneous metalwork assemblies.
M4 The lengths of *ladders* shall be measured along the lengths of stringers. The lengths of *handrails* and *bridge parapets* shall be measured along their top members.			**A2** Where *ladders* include safety loops, rest platforms or returned stringers, item descriptions shall so state.
M5 The lengths of *miscellaneous framing* shall be measured along the external perimeter of framing.			
		C2 Items for *plate* and *open grid flooring* shall be deemed to include supporting metalwork unless otherwise stated.	
		C3 Items for *welded mesh panelling* and *duct covers* shall be deemed to include supporting metalwork unless otherwise stated.	
		C4 Items for *tie rods* shall be deemed to include concrete, reinforcement and joints.	
		C5 Items for *bridge bearings* shall be deemed to include for grouting beneath plates.	**A3** Item descriptions for *bridge bearings* shall state the composition and materials of the bearing.
			A4 Item descriptions for *tanks* shall state the principal dimensions.

CLASS O: TIMBER

Includes: **Timber components and fittings**
Timber decking
Fittings and fastenings to timber components and decking

Excludes: **Formwork to concrete (included in class G)**
Timber piles (included in class P)
Timber sleepers (included in class S)
Timber supports in tunnels (included in class T)
Timber fencing (included in class X)
Carpentry and joinery in simple building works incidental to civil engineering works (included in class Z)

FIRST DIVISION		SECOND DIVISION	THIRD DIVISION
1 Hardwood components	m	**1** Cross-sectional area: not exceeding 0.01 m^2	**1** Length: not exceeding 1.5 m
2 Softwood components	m	**2** 0.01–0.02 m^2	**2** 1.5–3 m
		3 0.02–0.04 m^2	**3** 3–5 m
		4 0.04–0.1 m^2	**4** 5–8 m
		5 0.1–0.2 m^2	**5** 8–12 m
		6 0.2–0.4 m^2	**6** 12–20 m
		7 exceeding 0.4 m^2	**7** stated exceeding 20 m
3 Hardwood decking	m^2	**1** Thickness: not exceeding 25 mm	
4 Softwood decking	m^2	**2** 25–50 mm	
		3 50–75 mm	
		4 75–100 mm	
		5 100–125 mm	
		6 125–150 mm	
		7 exceeding 150 mm	
5 Fittings and fastenings	nr	**1** Straps	
		2 Spikes	
		3 Coach screws	
		4 Bolts	
		5 Plates	

MEASUREMENT RULES	DEFINITION RULES	COVERAGE RULES	ADDITIONAL DESCRIPTION RULES
		C1 Items for timber shall be deemed to include fixing, boring, cutting and jointing.	
M1 The length of timber *components* measured shall be their overall lengths with no allowance for scarfed or other joints.	**D1** The cross-sectional areas stated for classification of timber *components* shall be their nominal gross cross-sectional areas.		**A1** The nominal gross cross-sectional dimensions, thicknesses, grade or species and any impregnation requirements or special surface finishes shall be stated in item descriptions for timber *components*. **A2** The structural use and location of timber *components* shall be stated in item descriptions for components longer than 3 m.
M2 No deduction from the areas measured for timber *decking* shall be made for openings and holes each not exceeding 0.5 m^2 in area.	**D2** The thickness stated for classification of timber *decking* shall be the nominal gross thickness.		**A3** The nominal gross cross-sectional dimensions, thicknesses, species and any impregnation requirements or special surface finishes shall be stated in item descriptions for timber *decking*.
			A4 Materials, types and sizes of *fittings and fastenings* shall be stated in item descriptions.

CLASS P: PILES

Excludes: **Boring for site investigation (included in class B)**
Ground anchors (included in class C)
Walings and tie rods (included in class N)
Piling ancillaries (included in class Q)

FIRST DIVISION	SECOND DIVISION	THIRD DIVISION	
1 Bored cast in place concrete piles **2** Driven cast in place concrete piles	**1** Diameter: 300 mm or 350 mm **2** 400 mm or 450 mm **3** 500 mm or 550 mm **4** 600 mm or 750 mm **5** 900 mm or 1050 mm **6** 1200 mm or 1350 mm **7** 1500 mm	**1** Number of piles **2** Concreted length **3** Depth bored or driven to stated maximum depth	nr m m
3 Preformed concrete piles **4** Preformed concrete sheet piles **5** Timber piles	**1** Cross-sectional area: not exceeding 0.025 m^2 **2** $0.025–0.05 \text{ m}^2$ **3** $0.05–0.1 \text{ m}^2$ **4** $0.1–0.15 \text{ m}^2$ **5** $0.15–0.25 \text{ m}^2$ **6** $0.25–0.5 \text{ m}^2$ **7** $0.5–1 \text{ m}^2$ **8** exceeding 1 m^2	**1** Number of piles of stated length **2** Depth driven	nr m
6 Isolated steel piles	**1** Mass: not exceeding 15 kg/m **2** 15–30 kg/m **3** 30–60 kg/m **4** 60–120 kg/m **5** 120–250 kg/m **6** 250–500 kg/m **7** 500 kg/m – 1 t/m **8** exceeding 1 t/m	**1** Number of piles of stated length **2** Depth driven	nr m

MEASUREMENT RULES	DEFINITION RULES	COVERAGE RULES	ADDITIONAL DESCRIPTION RULES
M1 Bored and driven depths shall be measured along the axes of piles from the Commencing Surface to the toe levels of bored piles, to the bottom of the casings of driven cast in place piles and to the bottom of the toes of other driven piles. The Commencing Surface adopted in the preparation of the Bill of Quantities as the surface at which boring or driving is expected to begin shall be adopted for the measurement of the completed work.		**C1** Items for piles and stone columns shall be deemed to include disposal of excavated material.	**A1** Materials of which piles are composed shall be stated in item descriptions. **A2** Preliminary piles shall be identified in item descriptions. Raked piles shall be identified in item descriptions and their inclination ratios stated. **A3** The structure to be supported and the Commencing Surface shall be identified in item descriptions for piles.
M2 Each group of items for *cast in place concrete piles* shall comprise (*a*) an item for the *number of piles* (P 1–2 * 1) (*b*) an item for the total *concreted length* of piles (P 1–2 * 2) (*c*) an item for the total *depth bored or driven* (P 1–2 * 3). **M3** The *concreted lengths* of *cast in place concrete piles* shall be measured from the cut-off levels expressly required to the toe levels expressly required.	**D1** The maximum depth stated in item descriptions for the depth of *cast in place concrete piles* shall be the depth which is not exceeded by any pile included in the item. **D2** Piles comprising a driven permanent steel casing which is filled with concrete shall be classed as *driven cast in place concrete piles* where the piles are designed for the load to be carried on the concrete.		**A4** The diameter shall be stated in item descriptions for *cast in place concrete piles*. **A5** Contiguous bored piles shall be identified in item descriptions.
M4 Each group of items for *preformed concrete* and *timber piles* shall comprise (*a*) one or more items for the *number of piles of stated length* (P 3–6 * 1) (*b*) an item for the total *depth driven* (P 3–6 * 2).	**D3** The lengths of *preformed concrete* and *timber piles* shall be the lengths expressly required to be supplied excluding extensions but including heads and shoes.		**A6** The cross-section type and cross-sectional dimensions or diameter shall be stated in item descriptions for *preformed concrete* and *timber piles*. **A7** Item descriptions for *preformed concrete piles* shall state if they are prestressed. **A8** Details of treatments and coatings shall be stated in item descriptions for the *number of piles*. **A9** Details of driving heads and of shoes shall be stated in item descriptions for the *number of piles*.
M5 Each group of items for *isolated steel piles* shall comprise (*a*) one or more items for the *number of piles of stated length* (P 7 * 1) (*b*) an item for the total *depth driven* (P 7 * 2).	**D4** The lengths of *isolated steel piles* shall be the lengths expressly required to be supplied excluding extensions. **D5** Piles comprising a driven permanent steel casing which is filled with concrete shall be classed as *isolated steel piles* where the piles are designed for the load to be carried on the casing. Filling such piles shall be classed as *filling hollow piles with concrete* (Q 5 3 *).		**A10** The mass per metre and cross-sectional dimensions shall be stated in item descriptions for *isolated steel piles*. **A11** Details of treatments and coatings shall be stated in item descriptions for the *number of piles*.

CLASS P

FIRST DIVISION	SECOND DIVISION	THIRD DIVISION	
7 Interlocking steel piles	**1** Section modulus: not exceeding 500 cm^3/m	**1** Length of special piles	m
	2 500–800 cm^3/m	**2** Driven area	m^2
	3 800–1200 cm^3/m	**3** Area of piles of length: not exceeding 14 m	m^2
	4 1200–2000 cm^3/m	**4** 14–24 m	m^2
	5 2000–3000 cm^3/m	**5** exceeding 24 m	m^2
	6 3000–4000 cm^3/m		
	7 4000–5000 cm^3/m		
	8 exceeding 5000 cm^3/m		
8 Stone columns	**1** Diameter: not exceeding 300 mm	**1** Number of columns	nr
	2 300 mm – 450 mm	**2** Length of columns	m
	3 450 mm – 600 mm		
	4 600 mm – 750 mm		
	5 750 mm – 900 mm		
	6 900 mm – 1050 mm		
	7 1050 mm – 1200 mm		
	8 exceeding 1200 mm		

MEASUREMENT RULES	DEFINITION RULES	COVERAGE RULES	ADDITIONAL DESCRIPTION RULES
M6 Each group of items for *interlocking steel piles* shall comprise (*a*) one or more items for the total length of each type of *special pile*, if any (P 8 * 1) (*b*) an item for the total *driven area* of piles (P 8 * 2) (*c*) one or more items for the total *area of piles* divided into the ranges of length given in the third division (P 8 * 3–5). **M7** The areas of *interlocking steel piles* shall be calculated by multiplying the mean undeveloped horizontal lengths of the pile walls formed (including lengths occupied by special piles) by the depths measured in accordance with rule M1 in the case of items for the driven areas (P 8 * 2) and by the lengths defined in accordance with rule D6 in the case of items for the areas of piles (P 8 * 3–5). **M8** Closure and taper piles classed as *special piles* shall be measured only where they are expressly required.	**D6** The lengths of *interlocking steel piles* shall be the lengths expressly required to be supplied excluding extensions. **D7** Interlocking steel corner, junction, closure and taper piles shall be classed as *special piles*.		**A12** The section reference or mass per metre and section modulus shall be stated in item descriptions for *interlocking steel piles*. **A13** Details of treatments and coatings shall be stated in item descriptions for *area of piles*. **A14** The type of special pile shall be stated in item descriptions for the *length of special piles*.
		C2 Items for *stone columns* shall include for boring.	**A15** The diameter shall be stated in item descriptions for *stone columns*.

CLASS Q: PILING ANCILLARIES

Includes: **Work ancillary to piling**
Excludes: **Ground anchors (included in class C)**
Piles (included in class P)
Walings and tie rods (included in class N)

FIRST DIVISION	SECOND DIVISION		THIRD DIVISION
1 Cast in place concrete piles	**1** Pre-boring	m	**1** Diameter: 300 mm or 350 mm
	2 Backfilling empty bore with stated material	m	**2** 400 mm or 450 mm
	3 Permanent casings each length: not exceeding 13 m	m	**3** 500 mm or 550 mm
	4 exceeding 13 m	m	**4** 600 mm or 750 mm
	5 Enlarged bases	nr	**5** 900 mm or 1050 mm
	7 Cutting off surplus lengths	m	**6** 1200 mm or 1350 mm
	8 Preparing heads	nr	**7** 1500 mm
2 Cast in place concrete piles	**1** Reinforcement	t	**1** Straight bars, nominal size: not exceeding 25 mm
			2 exceeding 25 mm
			3 Helical bars of stated nominal size
3 Preformed concrete piles	**1** Pre-boring	m	**1** Cross-sectional area: not exceeding 0.025 m^2
4 Timber piles	**2** Jetting	m	**2** 0.025–0.05 m^2
	3 Filling hollow piles with concrete	m	**3** 0.05–0.1 m^2
	4 Number of pile extensions	nr	**4** 0.1–0.15 m^2
	5 Length of pile extensions, each length: not exceeding 3 m	m	**5** 0.15–0.25 m^2
	6 exceeding 3 m	m	**6** 0.25–0.5 m^2
	7 Cutting off surplus lengths	m	**7** 0.5–1 m^2
	8 Preparing heads	nr	**8** stated exceeding 1 m^2
5 Isolated steel piles	**1** Pre-boring	m	**1** Mass: not exceeding 15 kg/m
	2 Jetting	m	**2** 15–30 kg/m
	3 Filling hollow piles with concrete	m	**3** 30–60 kg/m
	4 Number of pile extensions	nr	**4** 60–120 kg/m
	5 Length of pile extensions, each length: not exceeding 3 m	m	**5** 120–250 kg/m
	6 exceeding 3 m	m	**6** 250–500 kg/m
	7 Cutting off surplus lengths	m	**7** 500 kg/m – 1 t/m
	8 Preparing heads	nr	**8** stated exceeding 1 t/m

MEASUREMENT RULES	DEFINITION RULES	COVERAGE RULES	ADDITIONAL DESCRIPTION RULES
M1 Work in this class, other than *backfilling empty bore* for cast in place concrete piles, shall be measured only where it is expressly required.		**C1** Items for piling ancillaries shall be deemed to include disposal of surplus materials unless otherwise stated.	
M2 The lengths of *permanent casings* shall be measured from the Commencing Surface to the bottom of the casing.	**D1** The *diameter* used for classification in the third division shall be the diameter of the piles.	**C2** Items for *permanent casings* shall be deemed to include driving heads and shoes.	**A1** The diameters of *enlarged bases* for bored piles shall be stated in item descriptions. **A2** Materials, thickness and details of treatments and coatings shall be stated in item descriptions for *permanent casings*. Item descriptions for *cutting off surplus lengths* which include permanent casings shall so state.
M3 The mass measured for *reinforcement* shall include the mass of reinforcement in laps. **M4** The mass of steel *reinforcement* shall be taken as 0.785 kg/m per 100 mm^2 of cross-section (7.85 t/m^3). The mass of other reinforcing materials shall be taken as stated in the Contract.	**D2** The nominal size used for classification in item descriptions for bar *reinforcement* shall be the cross-sectional size.	**C3** Items for *reinforcement* shall be deemed to include supporting reinforcement.	**A3** Materials shall be stated in item descriptions for *reinforcement*. **A4** Details of couplers for high tensile steel reinforcement which are expressly required shall be stated in item descriptions for *reinforcement*.
M5 Driving extended piles shall be included in the measurement of the items for driven depth in class P. **M6** Each group of items for *pile extensions* shall comprise (*a*) an item for the *number of pile extensions* (Q 3–4 4 *) (*b*) one or two items for the *length of pile extensions* divided into the ranges of length given in the second division (Q 3–4 5–6 *). **M7** The *length of pile extensions* measured shall not include lengths formed from material arising from cutting off surplus lengths of other piles. **M8** The length measured for timber pile extensions shall include lengths occupied by scarfed or other joints.	**D3** The *cross-sectional area* used for classification in the third division shall be the cross-sectional area of the piles.	**C4** Items for *pre-boring* shall be deemed to include grouting voids between the pile and the bore. **C5** Items for *pile extensions* shall be deemed to include the work necessary to attach the extension to the pile. **C6** Items for *filling hollow piles with concrete* shall be deemed to include removal of material from within the pile before concreting.	**A5** Item descriptions for *filling hollow piles with concrete* shall state the specification of the concrete. **A6** Materials of which *pile extensions* are composed shall be stated in item descriptions for their *length*.
M5 Driving extended piles shall be included in the measurement of the items for driven depth in class P. **M9** Each group of items for *pile extensions* shall comprise (*a*) an item for the *number of pile extensions* (Q 5 4 *) (*b*) one or two items for the *length of pile extensions* divided into the ranges of length given in the second division (Q 5 5–6 *). **M7** The *length of pile extensions* measured shall not include lengths formed from material arising from cutting off surplus lengths of other piles.	**D4** The *mass* used for classification in the third division shall be the mass of the piles.	**C4** Items for *pre-boring* shall be deemed to include grouting voids between the pile and the bore. **C5** Items for *pile extensions* shall be deemed to include the work necessary to attach the extension to the pile. **C6** Items for *filling hollow piles with concrete* shall be deemed to include removal of material from within the pile before concreting.	**A5** Item descriptions for *filling hollow piles with concrete* shall state the specification of the concrete. **A6** Materials of which *pile extensions* are composed shall be stated in item descriptions for their *length*.

CLASS Q

FIRST DIVISION		SECOND DIVISION		THIRD DIVISION
6 Interlocking steel piles		**1** Pre-boring	m	**1** Section modulus: not exceeding 500 cm^3/m
		2 Jetting	m	
		4 Number of pile extensions	nr	**2** 500–800 cm^3/m
		5 Length of pile extensions, each length: not exceeding 3 m	m	**3** 800–1200 cm^3/m
		6 exceeding 3 m	m	**4** 1200–2000 cm^3/m
		7 Cutting off surplus lengths	m	**5** 2000–3000 cm^3/m
		8 Preparing heads	nr	**6** 3000–4000 cm^3/m
				7 4000–5000 cm^3/m
				8 stated exceeding 5000 cm^3/m
7 Obstructions	h			
8 Pile tests	nr	**1** Maintained loading with various reactions		**1** Test load: not exceeding 100 t
		2 Constant rate of penetration		**2** 100–200 t
		3 Horizontal loading		**3** 200–300 t
				4 300–400 t
				5 400–600 t
				6 600–800 t
				7 800–1000 t
				8 exceeding 1000 t
		4 Non-destructive integrity		
		5 Inclinometer installations		

CLASS Q

MEASUREMENT RULES	DEFINITION RULES	COVERAGE RULES	ADDITIONAL DESCRIPTION RULES
M5 Driving extended piles shall be included in the measurement of the items for driven depth in class P. **M10** Each group of items for *pile extensions* shall comprise (*a*) an item for the *number of pile extensions* (Q 6 4 *) (*b*) one or two items for the *length of pile extensions* divided into the ranges of length given in the second division (Q 6 5–6 *). **M7** The *length of pile extensions* measured shall not include lengths formed from material arising from cutting off surplus lengths of other piles. **M11** The lengths measured for *cutting off surplus lengths* of interlocking steel piles shall be the mean undeveloped horizontal lengths to be cut (including lengths occupied by special piles).	**D5** The *section modulus* used for classification in the third division shall be the section modulus of the piles.	**C5** Items for *pile extensions* shall be deemed to include the work necessary to attach the extension to the pile.	**A6** Materials of which *pile extensions* are composed shall be stated in item descriptions for their *length*.
M12 *Obstructions* shall be measured only for breaking out rock or artificial hard material encountered above the founding stratum of bored piles.			
			A7 Item descriptions for *pile tests* shall identify those which are to preliminary piles. **A8** Item descriptions for loading tests shall state the load. Where the load is applied to raking piles item descriptions shall so state.

CLASS R: ROADS AND PAVINGS

Includes: **Sub-base, base and surfacing of roads, runways and other paved areas**
Kerbing and light duty pavements, footways and cycle tracks
Traffic signs and markings

Excludes: **Earthworks (included in class E)**
Drainage (included in classes I, J, K and L)
Fences and gates (included in class X)
Gantries and other substantial structures supporting traffic signs
Maintenance of roads and pavings

FIRST DIVISION	SECOND DIVISION		THIRD DIVISION
1 Unbound sub-base	**1** Type 1 unbound mixtures	m^2	**1** Depth: not exceeding 30 mm
	2 Type 2 unbound mixtures	m^2	**2** 30–60 mm
	3 Type 3 (open graded) unbound mixtures	m^2	**3** 60–100 mm
	4 Category B (close graded) unbound mixtures	m^2	**4** 100–150 mm
	5 Type 4 (asphalt arrisings) unbound mixtures	m^2	**5** 150–200 mm
			6 200–250 mm
			7 250–300 mm
			8 exceeding 300 mm
	6 Geotextiles	m^2	
	7 Additional depth of stated material	m^3	
2 Cement and other hydraulically bound pavements	**1** Cement bound granular mixture	m^2	**1** Depth: not exceeding 30 mm
	2 Fly ash bound mixture 1 and hydraulic road binder bound mixture 1	m^2	**2** 30–60 mm
	3 Slag bound mixture B2, fly ash bound mixture 2 and hydraulic road binder bound mixture 2	m^2	**3** 60–100 mm
	4 Slag bound mixture B3, fly ash bound mixture 3 and hydraulic road binder bound mixture 3	m^2	**4** 100–150 mm
	5 Fly ash bound mixture 5	m^2	**5** 150–200 mm
	6 Slag bound mixtures B1-1 B1-2 B1-3 B1-4	m^2	**6** 200–250 mm
	7 Soil treated by cement	m^2	**7** 250–300 mm
	8 Soil treated by other stated material	m^2	**8** exceeding 300 mm
3 Bituminous bound pavements	**1** Hot rolled asphalt base	m^2	
	2 Hot rolled asphalt binder	m^2	
	3 Dense base and binder course asphalt concrete with paving grade bitumen (recipe mixtures)	m^2	
	4 Dense asphalt concrete surface course	m^2	
	5 Hot rolled asphalt surface concrete	m^2	
	6 Close graded asphalt concrete surface course	m^2	
	7 Fine graded asphalt concrete surface course	m^2	
	8 Open graded asphalt concrete surface course	m^2	
4 Bituminous bound pavements	**1** Dense base and binder course asphalt concrete (design mixtures)	m^2	
	2 EME2 base and binder course asphalt concrete	m^2	
	3 Stone mastic asphalt binder course and regulating course	m^2	
	4 Porous asphalt	m^2	
	5 Hot rolled asphalt surface course and binder course	m^2	
	6 Cold recycled bound material	m^2	
	7 Cold milling/planning	m^2	
	8 Regulating course of stated material	t	

MEASUREMENT RULES	DEFINITION RULES	COVERAGE RULES	ADDITIONAL DESCRIPTION RULES
	D1 Types, categorized and mixtures refer to the Manual of Contract Documents for Highway Works, Volume 1 Specification for Highway Works dated November 2009 published by The Stationery Office.	**C1** Items for work in this class involving in situ concrete shall be deemed to include formwork and finishes to concrete.	**A1** Item descriptions for all courses of paving, road making materials and pavement slabs shall identify the material and state the depth of each course or slab and the spread rate of applied surface finishes. **A2** Item descriptions for work in this class which is applied to surfaces inclined at an angle exceeding 10° to the horizontal shall so state.
M1 The width of each course of materials shall be measured at the top surface of that course. The areas of manhole covers and other intrusions into a course shall not be deducted where the area of the intrusion is less than 1 m^2. **M2** The area of additional *geotextiles* in laps shall not be measured.			
			A3 The type and grade of material shall be stated in item descriptions for *geotextiles*.
			A4 The type of mixtures shall be stated in item descriptions for *cement and other hydraulically bound pavements*.
			A5 The type of mixtures shall be stated for *bituminous bound pavements*.
		C2 Items for *cold milling / planning* shall be deemed to include for disposal.	**A6** Item descriptions for *cold recycled bound materials* shall state whether it is in situ or ex situ.

CLASS R

FIRST DIVISION	SECOND DIVISION	THIRD DIVISION
5 Concrete pavements	**1** Unreinforced concrete surface slabs m² **2** Jointed reinforced concrete surface slabs m² **3** Continuously reinforced concrete surface slabs m² **4** Continuously reinforced concrete base slabs m² **5** Wet lean concrete m²	**1** Depth: not exceeding 30 mm **2** 30–60 mm **3** 60–100 mm **4** 100–150 mm **5** 150–200 mm **6** 200–250 mm **7** 250–300 mm **8** exceeding 300 mm
	6 Steel fabric reinforcement m²	**1** Nominal mass: not exceeding 2 kg/m² **2** 2–3 kg/m² **3** 3–4 kg/m² **4** 4–5 kg/m² **5** 5–6 kg/m² **6** 6–7 kg/m² **7** 7–8 kg/m² **8** stated exceeding 8 kg/m²
	7 Steel bar reinforcement t	**1** Nominal size: 6 mm **2** 8 mm **3** 10 mm **4** 12 mm **5** 16 mm **6** 20 mm **7** 25 mm **8** 32 mm or greater
	8 Separation and waterproof membranes m²	
6 Joints in concrete pavements m	**1** Transverse joints **2** Expansion joints **3** Contraction joints **4** Warping joints **5** Longitudinal joints **6** Construction joints	**1** Depth of joint: not exceeding 30 mm **2** 30–60 mm **3** 60–100 mm **4** 100–150 mm **5** 150–200 mm **6** 200–250 mm **7** 250–300 mm **8** exceeding 300 mm
7 Kerbs, channels, edgings, footways and paved areas	**1** Precast concrete kerbs **2** Precast concrete channels **3** Precast concrete edgings **4** Asphalt kerbs	**1** Straight or curved to radius exceeding 12 m m **2** Curved to radius not exceeding 12 m m **3** Quadrants nr **4** Drops nr **5** Transitions nr
	5 Precast concrete, natural stone, block and clay slabs and pavers m² **6** Flexible surfacing m² **7** In situ concrete surfacing m² **8** Grass concrete paving m²	
8 Ancillaries	**1** Traffic signs	**1** Non-illuminated nr **2** Illuminated nr
	2 Surface markings	**1** Non-reflecting road studs nr **2** Reflecting road studs nr **3** Letters and shapes nr **4** Continuous lines m **5** Intermittent lines m **6** Raised rib m

MEASUREMENT RULES	DEFINITION RULES	COVERAGE RULES	ADDITIONAL DESCRIPTION RULES
			A7 Item descriptions for *wet lean concrete* shall state the mix.
M3 The areas of additional *fabric reinforcement* in laps shall not be measured. **M4** The mass of steel *reinforcement* shall be taken as 0.785 kg/m per 100 mm^2 of cross-section (7.85 t/m^3). The mass of other reinforcing materials shall be taken as stated in the Contract. **M5** The mass of *reinforcement* measured shall include the mass of steel supports to top reinforcement.		**C3** Items for *reinforcement* shall be deemed to include supporting reinforcement other than steel supports to top reinforcement.	**A8** Item descriptions for *steel fabric reinforcement* or size and mass per square metre. **A9** Item descriptions for *steel bar reinforcements* shall state the type of reinforcement.
M6 The areas of additional *in separation and waterproof membranes* in laps shall not be measured.			**A10** Item descriptions for *separation and waterproof membranes* shall state their materials and thickness.
M7 *Construction joints* shall be measured only where they are at locations where construction joints are expressly required.			**A11** The dimensions, spacing and nature of sealed grooves and rebates, waterstops, dowels and other components shall be stated in item descriptions for *joints in concrete pavements.*
M8 The width of each course of materials shall be measured at the top surface of that course. The areas of manhole covers and other intrusions into a course shall not be deducted where the area of the intrusion is less than 1 m^2.		**C4** Items for *kerbs, channels and edgings* shall be deemed to include beds, backings, reinforcement, joints and cutting. **C5** Item descriptions for *kerbs, channels and edgings* shall be deemed to include for earthworks and concrete ancillaries. **C6** Items for *precast concrete natural stone block and clay slabs and pavers* shall be deemed to include for beds, joints and cutting. **C7** Items for *flexible surfacing and in situ concrete surfacing* shall be deemed to include for beds and joints. **C8** Items for *grass concrete pavings* shall be deemed to include for beds, joints, soiling and seeding.	**A12** Materials and cross-sectional dimensions of *kerbs, channels and edgings* and their beds and backings shall be stated in item descriptions **A13** The material, size and thickness for *precast concrete natural stone block and clay slabs and pavers* shall be stated in item descriptions. **A14** The materials and thicknesses for *flexible surfacing in situ concrete surfacing and grass concrete pavings* shall be stated in item descriptions.
M9 Lengths measured for linear *surface markings* shall exclude gaps in intermittent markings. **M10** Items for support gantries and other substantial structures associated with *traffic signs* which are constructed in concrete, structural metalwork or other materials shall be given in the appropriate classes.		**C9** Items for *traffic signs* other than traffic signs measured in accordance with rule M10 shall be deemed to include foundations, supporting posts, excavation, preparation of surfaces, disposal of excavated material, removal of existing services, upholding sides of excavation, backfilling, concrete, reinforcement and joints.	**A15** The material, size and type shall be stated in item descriptions for *traffic signs* and *surface markings.* **A16** The shape and colour of aspects shall be stated in item descriptions for *reflecting road studs.*

CLASS S: RAIL TRACK

Includes: **Track foundations, rails, sleepers, fittings, switches and crossings**
Excludes: **Overhead crane rails (included in class M)**
Concrete track foundations (included in classes F and G)

FIRST DIVISION	SECOND DIVISION	THIRD DIVISION
1 Track foundations	**1** Bottom ballast m^3 **2** Top ballast m^3 **3** Blinding m^2 **4** Blankets and vibration mats m^2 **5** Waterproof membranes m^2 **6** Ballast cleaning m^3 **7** Tamping m **8** Pneumatic ballast injection: stone blowing m	
2 Taking up	**1** Bullhead rails **2** Flat bottom rails **3** Dock and crane rails	**1** Plain track m **2** Turnouts nr **3** Diamond crossings nr
	4 Check and guard rails m **5** Conductor rails m	
	6 Sundries nr	**1** Buffer stops nr **2** Retarders nr **3** Wheel stops nr **4** Lubricators nr **5** Switch heaters nr **6** Switch levers nr **7** Eutectic strip m
3 Lifting, packing and slewing and works to existing track nr	**1** Bullhead rail track **2** Flat bottom rail track **3** Spot replacement of sleepers **4** Existing rail turning and refixing **5** Rerailing **6** Stressing rail	**1** Plane track m **2** Plane track: with conductor rail m **3** Plain track: with turnout nr **4** Plain track: with conductor rail and turnout nr **5** Buffer stops

CLASS S

MEASUREMENT RULES	DEFINITION RULES	COVERAGE RULES	ADDITIONAL DESCRIPTION RULES
M1 The volume of *top ballast* measured shall include the volume occupied by sleepers. **M2** The areas of additional *waterproof membranes* in laps shall not be measured. **M3** Item descriptions for *ballast cleaning* shall state the depth of material to be cleaned and the cleaning material to be used to fill voids. **M4** Item description for *tamping* shall state the width of the ballast to be tamped. **M5** Item description for *pneumatic ballast injection* shall state the width of the ballast to be treated.	**D1** *Bottom ballast* shall be ballast placed before the track is laid. **D2** *Top ballast* shall be ballast placed after the track is laid.	**C1** Items for ballast cleaning shall be deemed to include the provision of ballast to fill voids.	**A1** Item descriptions for *track foundations* shall state the type of material. **A2** Item descriptions for *blinding, blankets, vibration mats* and *waterproof membranes* shall state their material and thickness.
M6 The length of *taking up plain track* shall be measured along the centre line of the track (two rail) and shall exclude lengths occupied by turnouts and diamond crossings. **M7** The lengths of *taking up check, guard* and *conductor* rails shall be measured along the lengths of the rail (one rail) and shall exclude lengths within turnouts and diamond crossings.		**C2** Items for *taking up turnouts* and *diamond crossings* shall be deemed to include check, guard and conductor rails.	**A3** Item descriptions for *taking up track* shall state the amount of dismantling, details of disposal of the track and the type of rail, sleeper and joint. **A4** Item descriptions for *taking up buffer stops* shall state their approximate weight and type of construction.
M8 The length of track stated in item descriptions for *lifting, packing and slewing track* shall be the length measured along the centre line of the track (two rail) and shall be taken over all roads. Switch roads shall be measured from the toes of switches. **M9** The length track for *rerailing stressing rail* shall be the length of a single rail and exclude length of turnouts and crossings. **M10** Item descriptions for spot replacement of sleepers shall state the type of rail fixings required and the method of fixing to track and to the sleeper.	**D3** Spot replacement of sleepers shall cover the replacement of sleepers in track neither supplied nor laid by the contractor.	**C3** Items for *lifting, packing and slewing* shall be deemed to include opening out, packing and boxing in with ballast and insertion of closure rails. **C4** Items for *spot replacement of sleepers* shall be deemed to include opening out of ballast, disposal of existing sleeper and replacing ballast. **C5** Items for *rerailing* shall be deemed to include for grinding, milling and planning. **C6** Items for spot replacement of sleepers shall be deemed to include opening out of ballast removing old sleeper, the supply and laying of new sleepers and repacking ballast.	**A5** Item descriptions for *lifting, packing and slewing* shall state the length of track, the maximum distance of slew and the maximum lift. **A6** Where extra ballast is required item descriptions for *lifting, packing and slewing* shall so state. **A7** Items for *spot replacement of sleepers* shall state the type of sleeper and type of fixing. **A8** Items for *existing rail turning and refixing* shall state the type of rail and type of fixing. **A9** Items for *rerailing* shall state the type of joints. **A10** Items for *stressing rail* shall state when the work is restressing.

CLASS S

FIRST DIVISION	SECOND DIVISION		THIRD DIVISION	
4 Supplying	**1** Bullhead rails	t	**1** Mass: not exceeding 20 kg/m	
	2 Flat bottom rails	t	**2** 20–30 kg/m	
	3 Dock and crane rails	t	**3** 30–40 kg/m	
	4 Check and guard rails	t	**4** 40–50 kg/m	
	5 Conductor rails	t	**5** exceeding 50 kg/m	
	6 Twist rails	nr		
	7 Sleepers	nr	**1** Timber	
			2 Concrete	
			3 Steel	
	8 Fittings	nr	**1** Chairs	
			2 Baseplates	
			3 Pandrol rail fastenings	
			4 Plain fishplates	
			5 Insulated fishplates	
			6 Conductor rail insulators	
			7 Conductor rail side ramps	
5 Supplying	**1** Turnouts and crossings	nr	**1** Trap points	nr
			2 Catch points	nr
			3 Turnouts	nr
			4 Diamond crossings	nr
	2 Sundries		**1** Buffer stops	nr
			2 Retarders	nr
			3 Wheel stops	nr
			4 Lubricators	nr
			5 Switch heaters	nr
			6 Switch levers	nr
			7 Conductor rail guard boards	m
	3 Prefabricated track panels 18.288 m long with rails and bearers	m	**1** Bullhead track: timber sleepers	
			2 concrete sleepers	
			3 steel sleepers	
			4 Flat bottom track: timber sleepers	
			5 concrete sleepers	
			6 steel sleepers	
	4 Prefabricated turnouts and crossings	nr	**1** Turnouts: type stated timber bearers	
			2 Turnouts: type stated concrete bearers	
			3 Turnouts: type stated steel bearers	
			4 Diamond crossings: type stated timber bearers	
			5 Diamond crossings: type stated concrete bearers	
			6 Diamond crossings: type stated steel bearers	
6 Laying: with bearers on ballast	**1** Bullhead rails		**1** Plain track	m
	2 Flat bottom rails		**2** Form curve in plain track radius not exceeding 300 m	m
	3 Dock rails		**3** Form curve in plain track radius exceeding 300 m	m
	4 Crane rails		**4** Turnouts	nr
			5 Diamond crossings	nr
			6 Welded joints	nr
	5 Check rails		**1** Rail	m
	6 Guard rails		**2** Length ends	nr
	7 Conductor rails		**3** Side ramps	nr
			4 Welded joints	nr

MEASUREMENT RULES	DEFINITION RULES	COVERAGE RULES	ADDITIONAL DESCRIPTION RULES
M11 The mass measured for *supplying rails* shall include the mass of twist rails. **M12** *Fishplates* shall be measured by the number of pairs.		**C7** Items for *supplying* shall be deemed to include delivery of components to the Site. **C8** Items for supplying *sleepers* shall be deemed to include fittings attached by the supplier. **C9** Items for supplying *fittings* shall be deemed to include fixings, keys, clips, bolts, nuts, screws, spikes, ferrules, track circuit insulators, pads and conductor rail insulator packings.	**A11** Item descriptions for supplying *rails* shall state either the section reference and the mass per metre or the cross-sectional dimension and the mass per metre. **A12** Item descriptions for supplying *sleepers* and *fittings* shall state the type. **A13** Hollow sleepers shall be identified in item descriptions for *sleepers.* **A14** Items for supplying *sleepers* shall state the size of the sleepers and identify the fittings which are attached by the supplier. Sleepers with integral ducts for services shall be so described.
M13 *Conductor rail guard boards* shall be measured each side of the rail.		**C10** Items for *supplying* shall be deemed to include delivery of components to the Site. **C11** Items for supplying *turnouts* and *diamond crossings* shall be deemed to include timbers, fittings and check rails. **C12** Items for supplying *conductor rail guard boards* shall be deemed to include fixings.	**A15** Item descriptions for supplying *prefabricated track panels* shall state the length of panels. **A16** Item descriptions for supplying *turnouts* and *crossings* shall state the type. **A17** Item descriptions for supplying *sundries* shall state the type. **A18** Item descriptions for supplying *buffer stops* shall state their approximate weight. **A19** Item descriptions for supplying *buffer stops* shall state the type and approximate weight.
M14 The length of *laying plain track* shall be measured along the centre line of the track (two rail) and shall include the lengths occupied by turnouts and diamond crossings. **M15** The lengths of *laying check, guard* and *conductor rails* shall be measured along the lengths of the rail (one rail). **M16** *Conductor rail guard boards* shall be measured each side of the rail. **M17** Laying rails direct to concrete bases shall be measured separately.	**D4** Laying rails direct to concrete bases shall be classed as slab track.	**C13** Items for *laying* shall be deemed to include work carried out after delivery of components to the Site or, where track is not to be supplied by the Contractor, to the location stated in accordance with rule A14. **C14** Laying rails direct to concrete, slab track, shall be deemed to include fixings and pads.	**A20** Item descriptions for *laying* rail track not supplied by the contractor shall indentify the form in which it is to be supplied. **A21** Item descriptions for *laying prefabricated track* shall identify pre-fabricated lengths. **A22** Item descriptions for *laying rails* shall state the type and mass per metre of rail and the type of joint and sleeper. **A23** Item descriptions for *laying prefabricated track panels* shall state the panel length. **A24** Item descriptions for laying *turnouts and diamond crossings* shall state their type and weight. **A25** Item descriptions for *welded joints* shall state the rail section and the type of weld. **A26** Item descriptions for *laying buffer stops* shall state their approximate weight.

CLASS S

FIRST DIVISION	SECOND DIVISION	THIRD DIVISION	
7 Laying	**1** Bullhead rails: slab track **2** Flat bottom rails: slab track	**1** Plain track	m
		2 Form curve in plain track radius not exceeding 300 m	m
		3 Form curve in plain track radius exceeding 300 m	m
		4 Turnouts	nr
		5 Diamond crossings	nr
		6 Welded joints	nr
	3 Prefabricated track panels	**1** Bullhead rail: timber sleepers	m
		2 Bullhead rail: concrete sleepers	m
		3 Flat bottom rail: timber sleepers	m
		4 Flat bottom rail: concrete sleepers	m
		5 Flat bottom rail: steel sleepers	m
	4 Prefabricated turnouts and crossings: with bearers	**1** Turnouts: timber bearers	nr
		2 Turnouts: concrete bearers	nr
		3 Turnouts: steel bearers	nr
		4 Diamond crossings: timber bearers	nr
		5 Diamond crossings: concrete bearers	nr
		6 Diamond crossings: steel bearers	nr
	5 Sundries	**1** Buffer stops	nr
		2 Retarders	nr
		3 Wheel stops	nr
		4 Lubricators	nr
		5 Switch heaters	nr
		6 Switch levers	nr
		7 Conductor rail guard boards	m
		8 Static sander	nr
		9 Eutectic strip	m

MEASUREMENT RULES	DEFINITION RULES	COVERAGE RULES	ADDITIONAL DESCRIPTION RULES
M14 The length of *laying plain track* shall be measured along the centre line of the track (two rail) and shall include the lengths occupied by turnouts and diamond crossings. **M15** The lengths of *laying check, guard* and *conductor rails* shall be measured along the lengths of the rail (one rail). **M16** *Conductor rail guard boards* shall be measured each side of the rail. **M17** Laying rails direct to concrete bases shall be measured separately.	**D4** Laying rails direct to concrete bases shall be classed as slab track.	**C13** Items for *laying* shall be deemed to include work carried out after delivery of components to the Site or, where track is not to be supplied by the Contractor, to the location stated in accordance with rule A14. **C14** Laying rails direct to concrete, slab track, shall be deemed to include fixings and pads.	**A20** Item descriptions for *laying* rail track not supplied by the contractor shall indentify the form in which it is to be supplied. **A21** Item descriptions for *laying prefabricated track* shall identify pre-fabricated lengths. **A22** Item descriptions for *laying rails* shall state the type and mass per metre of rail and the type of joint and sleeper. **A23** Item descriptions for *laying prefabricated track panels* shall state the panel length. **A24** Item descriptions for laying *turnouts and diamond crossings* shall state their type and weight. **A25** Item descriptions for *welded joints* shall state the rail section and the type of weld. **A26** Item descriptions for *laying buffer stops* shall state their approximate weight.

CLASS T: TUNNELS

Includes: **Excavation, lining and securing of tunnels, shafts and other subterranean cavities**
Excludes: **Geotechnical processes carried out from the ground surface (included in class C)**
Filling within tunnels (included in class E)
Reinforcement in in situ lining (included in class G)
Pipe laying in headings, tunnels and shafts (included in classes I, J, K and L)
Cut and cover tunnels

FIRST DIVISION	SECOND DIVISION		THIRD DIVISION	
1 Excavation	**1** Tunnels in rock	m^3	**1** Stated diameter:	not exceeding 2 m
	2 Tunnels in other stated material	m^3	**2**	2–3 m
	3 Shafts in rock	m^3	**3**	3–4 m
	4 Shafts in other stated material	m^3	**4**	4–5 m
	5 Other cavities in rock	m^3	**5**	5–6 m
	6 Other cavities in other stated material	m^3	**6**	6–7 m
			7	7–8 m
			8	exceeding 8 m
	7 Excavated surfaces in rock	m^2		
	8 Excavated surfaces in other stated material	m^2		

MEASUREMENT RULES	DEFINITION RULES	COVERAGE RULES	ADDITIONAL DESCRIPTION RULES
M1 Tunnels constructed by cut and cover are excluded from this class. The earthworks, in situ concrete and other components of tunnels constructed by cut and cover shall be classed appropriately.	**D1** Transitions, breakaways and intersections between tunnels and shafts which include work outside the normal profiles of the tunnels and shafts shall be classed as *other cavities*.		**A1** Where tunnelling work is expressly required to be executed under compressed air, items shall be so described. Item descriptions shall state the gauge pressure in stages. The first stage shall be gauge pressure not exceeding 1 bar. Subsequent stages shall be gauge pressures in increments of 0.4 bars. The provision and operation of plant and services associated with the use of compressed air shall be classed as *specified requirements* in class A. **A2** Item descriptions for excavation and lining of *other cavities* shall identify the cavity.
M2 The volume measured for *excavation* shall be calculated to the payment lines shown on the Drawings or, where no payment lines are shown, to the net dimensions of the volumes to be excavated. Excavation (other than overbreak) outside the normal cross-sectional profile of tunnels and shafts shall be classed as *excavation of other cavities*. **M3** An isolated volume of *rock* occurring within other material to be excavated shall not be measured separately unless its volume exceeds 0.25 m^3. **M4** The area measured for *excavated surfaces* shall be the area of the payment surfaces shown on the Drawings or, where no payment surfaces are shown, the net areas of the surfaces of the volumes to be excavated.	**D2** The diameter used for classification and stated in item descriptions shall be the external diameter of the excavation cross-section of *tunnels, shafts* and *other cavities*.	**C1** Items for *excavation* shall be deemed to include disposal of excavated material off the Site and removal of dead services unless otherwise stated in item descriptions.	**A3** Item descriptions for *excavation* shall state whether tunnels and shafts are straight, curved or tapered. Item descriptions for *excavation* shall state the gradient of tunnels sloping at a gradient of 1 in 25 or steeper and the inclination to the vertical of inclined shafts. **A4** Where material is for disposal on the Site the location of the disposal areas shall be stated in item descriptions for *excavation*. Where excavated material is to be used as filling, item descriptions shall so state. **A5** Details of filling for voids caused by overbreak shall be stated in item descriptions for *excavated surfaces*. **A6** Where *tunnels, shafts* and *other cavities* are not of circular cross-section, their maximum external dimension of cross-section shall be substituted for the diameter and their external cross-sectional dimensions shall be stated in item descriptions.

CLASS T

FIRST DIVISION	SECOND DIVISION		THIRD DIVISION		
2 In situ lining to tunnels **3** In situ lining to shafts **4** In situ lining to other cavities	**1** Sprayed concrete primary **2** Sprayed concrete secondary **3** Cast concrete primary **4** Cast concrete secondary **5** Formwork to stated finish	m^2 m^2 m^3 m^3 m^2	**1** Stated diameter: **2** **3** **4** **5** **6** **7** **8**	not exceeding 2 m 2–3 m 3–4 m 4–5 m 5–6 m 6–7 m 7–8 m exceeding 8 m	
5 Preformed segmental lining to tunnels **6** Preformed segmental lining to shafts **7** Preformed segmental lining to other cavities	**1** Precast concrete bolted rings **2** Precast concrete expanded rings **3** Cast iron bolted rings **4** Cast iron expanded rings **5** Nodular iron rings **6** Fabricated steel rings	nr nr nr nr nr nr			
	7 Lining ancillaries		**1** Parallel circumferential packing **2** Tapered circumferential packing **3** Stepped junctions **4** Caulking of stated material		nr nr nr m

MEASUREMENT RULES	DEFINITION RULES	COVERAGE RULES	ADDITIONAL DESCRIPTION RULES
M5 The thickness of *in situ lining* shall be measured to the payment lines shown on the Drawings or, where no payment lines are shown, to the net dimensions of the volumes to be lined (see rule M2). **M6** The volume of in situ cast concrete lining shall be calculated as set out in class F. **M7** The measurement unit for *packing* shall be the number of rings of segments packed.	**D3** Reinforcing materials added to the mix for *sprayed concrete* shall not be classed as reinforcement. **D4** The diameter of *lining* used for classification and stated in item descriptions shall be the internal diameter.	**C2** Items for *lining* shall be deemed to include joints and finishes. **C3** Items for *preformed lining* shall be deemed to include reinforcement and formwork.	**A7** Item descriptions for *lining* shall state whether tunnels and shafts are straight, curved or tapered. Item descriptions for *lining* shall state the gradient of tunnels sloping at a gradient of 1 in 25 or steeper and the inclination to the vertical of inclined shafts. **A8** Item descriptions for *in situ lining* shall state the specification of the concrete and whether it is reinforced and identify those linings which are to form head walls, shaft bottoms and other similar components. Item descriptions for *sprayed concrete lining* shall state the minimum thickness. **A9** Item descriptions for *preformed segmental lining rings* shall identify the components of each ring and state the nominal ring width and the maximum piece weight. **A10** Item descriptions for *preformed segmental lining* in pilot tunnels or shafts shall so state. Where the materials used in preformed segmental lining to pilot tunnels and shafts are to remain the property of the employer item descriptions shall so state. **A11** Item descriptions for *precast concrete segmental lining rings* shall state whether the segments are flanged or solid. Item descriptions for metal *segmental lining rings* which have machined abutting surfaces shall so state. **A12** The internal diameter of the *lining* shall be stated in item descriptions for *lining ancillaries*. **A13** Where *linings* are not of circular cross-section their maximum internal dimension of cross-section shall be substituted for the diameter and their internal cross-sectional dimensions shall be stated in item descriptions.

CLASS T

FIRST DIVISION	SECOND DIVISION		THIRD DIVISION	
8 Support and stabilization	**1** Rock bolts	m	**1** Mechanical **2** Mechanical grouted **3** Pre-grouted impacted **4** Chemical end anchor **5** Chemical grouted **6** Chemically filled	
	2 Internal support		**1** Steel arches: supply **2** erection **3** Timber supports: supply **4** erection **5** Lagging **6** Sprayed concrete **7** Mesh or link	t t m^3 m^3 m^2 m^2 m^2
	3 Pressure grouting		**1** Sets of drilling and grouting plant **2** Face packers **3** Deep packers of stated size **4** Drilling and flushing to stated diameter **5** Re-drilling and flushing **6** Injection of grout materials of stated composition	nr nr nr m m t
	4 Forward probing	m		

NOTE

Concrete work in *lining* to tunnels, shafts and other cavities which involves other than simple shapes may be classed as concrete (class F) and concrete ancillaries (class G).

MEASUREMENT RULES	DEFINITION RULES	COVERAGE RULES	ADDITIONAL DESCRIPTION RULES
M8 Both temporary and permanent *support and stabilization* shall be measured. **M9** The mass measured for *steel arches* shall be calculated as set out in class M. The volume of *timber support* measured shall be the volume of support in timber components calculated as set out in class O. The area of *sprayed concrete support* shall be measured at the payment lines shown on the Drawings or, where no payment lines are shown, to the net dimensions of the support to be provided. **M10** The number of *face packers* and *deep packers* measured shall be the number of injections. **M11** The mass measured for *injection of grout materials* shall not include the mass of mixing water.	**D3** Reinforcing materials added to the mix for *sprayed concrete* shall not be classed as reinforcement. **D5** Mesh or link reinforcement in *sprayed concrete support* shall be classed as mesh or link support.	**C4** Items for *face packers* shall be deemed to include collaring, securing and making good linings on completion.	**A14** Item descriptions for *rock bolts* shall state their size, type, shank detail and maximum length. **A15** The materials used for lagging and for packing or grouting behind lagging shall be stated in item descriptions. **A16** Item descriptions for *sprayed concrete support* shall state the specification of the concrete and whether it is reinforced and the minimum thickness. Item descriptions for *mesh or link support* shall state the size and mass of mesh or link fabric. **A17** The lengths of holes shall be stated in stages of 5 m in item descriptions for *drilling* and *re-drilling* holes for pressure grouting and in item descriptions for *forward probing*.

CLASS U: BRICKWORK, BLOCKWORK AND MASONRY

Excludes: Brickwork in manholes and other brickwork incidental to pipework (included in class K)
Brickwork in sewer renovation (included in class Y)

FIRST DIVISION	SECOND DIVISION	THIRD DIVISION
1 Common brickwork **2** Facing brickwork **3** Engineering brickwork **4** Lightweight blockwork **5** Dense concrete blockwork **6** Artificial stone blockwork **7** Ashlar masonry **8** Rubble masonry	**1** Thickness: not exceeding 150 mm m^2 **2** 150–250 mm m^2 **3** 250–500 mm m^2 **4** 500 mm–1 m m^2 **5** exceeding 1 m m^2	**1** Vertical straight walls **2** Vertical curved walls **3** Battered straight walls **4** Battered curved walls **5** Vertical facing to concrete **6** Battered facing to concrete **7** Casing to metal sections
	6 Columns and piers of stated cross-sectional dimensions m	
	7 Surface features	**1** Copings and sills, material stated m **2** Rebates and chases m **3** Cornices m **4** Band courses m **5** Corbels m **6** Pilasters m **7** Plinths m **8** Fair facing m^2

MEASUREMENT RULES	DEFINITION RULES	COVERAGE RULES	ADDITIONAL DESCRIPTION RULES
M1 Each skin of brickwork, blockwork or masonry which is in cavity or composite construction shall be measured. **M2** Volumes and areas measured for brickwork, blockwork and masonry shall include the volumes and areas of joints, and exclude those of copings and sills. No deduction from or addition to volumes and areas measured shall be made for intruding or projecting surface features. No deduction from the volumes and areas measured shall be made for holes and openings in walls or surfaces each not exceeding 0.25 m^2 in cross-sectional area.			
M3 Mean dimensions shall be used to calculate the areas and volumes of *walls, facing to concrete* and *casing to metal sections* and the heights of *columns and piers*.	**D1** Walls or facing battered on one or both sides shall be classed as *battered walls* or *battered facing*. **D2** Isolated walls having a length on plan not exceeding four times their thickness shall be classed as *piers*. **D3** In determining the thicknesses of *walls, facing* and *casing*, the presence of surface features shall be ignored. **D4** The thicknesses of *battered walls* shall be their mean thicknesses.	**C1** Items for *masonry* shall be deemed to include fair facing.	**A1** Item descriptions for *walls, facing to concrete, casing to metal sections, columns and piers* shall either state the materials, nominal dimensions and types of brick, block and stone or give equivalent references to applicable British Standard specifications. **A2** Item descriptions for *masonry walls, facing to concrete, casing to metal sections, columns and piers* shall state the surface finish. **A3** The bonding pattern, type of mortar and type of jointing and pointing shall be stated in item descriptions for *walls, facing to concrete, casing to metal sections, columns and piers*. **A4** Item descriptions for *walls, facing to concrete, casing to metal sections, columns and piers* which are in cavity or composite construction shall so state. **A5** The nominal thickness of *walls, facing to concrete* and *casing to metal sections* shall be stated in item descriptions.
M4 The lengths of *surface features* measured shall be mean lengths. The areas measured for *fair facing* shall be those expressly required and shall be measured at the face.	**D5** Columns and piers attached to walls or facing of the same material shall be classed as *pilasters*.		**A6** Item descriptions for *surface features* shall include sufficient detail to identify special masonry and special or cut bricks and blocks. The spacing of intermittent surface features shall be stated in item descriptions. **A7** The cross-sectional dimensions of *surface features* shall be stated in item descriptions where the cross-sectional area of the surface feature exceeds 0.05 m^2.

CLASS U

FIRST DIVISION	SECOND DIVISION	THIRD DIVISION	
	8 Ancillaries	**1** Joint reinforcement	m
		2 Damp proof courses	m
		3 Movement joints	m
		4 Bonds to existing work	m^2
		5 Infills of stated thickness	m^2
		6 Fixings and ties	m^2
		7 Built-in pipes and ducts, cross-sectional area: not exceeding 0.05 m^2	nr
		8 stated exceeding 0.05 m^2	nr

MEASUREMENT RULES	DEFINITION RULES	COVERAGE RULES	ADDITIONAL DESCRIPTION RULES
M5 The length of additional material in laps of *joint reinforcement* and *damp proof courses* shall not be measured. **M6** The areas of *fixings and ties* measured shall be the areas of brickwork, blockwork or masonry fixed or tied. Where two areas of brickwork, blockwork or masonry are fixed or tied to each other, only the smaller of the two areas shall be measured.		**C2** Items for *built-in pipes and ducts* shall be deemed to include their supply unless otherwise stated.	**A8** The materials and dimensions of *joint reinforcement* and *damp proof courses*, the materials of *infills* and the type and spacing of *fixings and ties* shall be stated in item descriptions. **A9** The dimensions and nature of components including face or internal details shall be stated in item descriptions for *movement joints*. **A10** The lengths of built-in pipes and ducts shall be stated in item descriptions where they exceed 1 m.

CLASS V: PAINTING

Includes: **In situ painting**
Excludes: **Painting carried out prior to delivery of components to the Site**

FIRST DIVISION	SECOND DIVISION	THIRD DIVISION
1 Iron or zinc based primer paint **2** Etch primer paint **3** Oil paint **4** Alkyd gloss paint **5** Emulsion paint **6** Cement paint **7** Epoxy or polyurethane paint **8** Bituminous or coal tar paint	**1** Metal, other than metal sections and pipework **2** Timber **3** Smooth concrete **4** Rough concrete **5** Masonry **6** Brickwork and blockwork	**1** Upper surfaces inclined at an angle not exceeding 30° to the horizontal m^2 **2** Upper surfaces inclined at 30°–60° to the horizontal m^2 **3** Surfaces inclined at an angle exceeding 60° to the horizontal m^2 **4** Soffit surfaces and lower surfaces inclined at an angle not exceeding 60° to the horizontal m^2 **6** Surfaces of width not exceeding 300 mm m **7** Surfaces of width 300 mm–1 m m
		8 Isolated groups of surfaces nr
	7 Metal sections m^2 **8** Pipework m^2	

NOTE

Painting may be measured by the number of *isolated groups of surfaces* of the same shape and dimensions instead of by the length or area of the separate surfaces.

CLASS V

MEASUREMENT RULES	DEFINITION RULES	COVERAGE RULES	ADDITIONAL DESCRIPTION RULES
M1 No deduction from the areas measured shall be made for holes and openings in the painted surfaces each not exceeding 0.5 m^2 in area.		**C1** Items for painting shall be deemed to include preparation of surfaces before painting.	**A1** Item descriptions for painting shall state the material used and either the number of coats or the film thickness. **A2** Preparation of surfaces shall be identified in item descriptions where more than one type of preparation is specified for the same surface.
M2 Surfaces of width not exceeding 1 m shall not be distinguished by inclination.			
M3 *Isolated groups of surfaces* of different shape or dimensions shall be measured as separate items.	**D1** *Isolated groups of surfaces* shall be classed as such only where the total surface area of each group does not exceed 6 m^2.		**A3** Item descriptions for *isolated groups of surfaces* shall identify the work to be painted and its location.
M4 In calculating the painted area of *metal sections* the presence of connecting plates, brackets, rivets, bolts, nuts and similar projections shall be ignored. **M5** The area measured for painting *pipework* shall be the length multiplied by the girth of each length of pipe or lagged pipe with no deductions or additions for flanges, valves, other projecting fittings and hangers.		**C2** Painting of *metal sections* shall be deemed to include painting the surfaces of connecting plates, brackets, rivets, bolts, nuts and similar projections. **C3** Painting of *pipework* shall be deemed to include painting the surfaces of flanges, valves, other projecting fittings and hangers.	

CLASS W: WATERPROOFING

Includes: **Damp proofing, tanking and roofing**
Excludes: **Waterproofed joints (included in classes C, G, H, I, J, K, R, T, U, X and Y)**
Damp proof courses in brickwork, blockwork and masonry (included in class U)
Surface finishes and linings in simple building works incidental to civil engineering works (included in class Z)

FIRST DIVISION	SECOND DIVISION	THIRD DIVISION
1 Damp proofing **2** Tanking **3** Roofing	**1** Asphalt **2** Sheet metal **3** Waterproof sheeting **4** Waterproof coating **5** Rendering in ordinary cement mortar **6** Rendering in waterproof cement mortar **7** Tiles	**1** Upper surfaces inclined at an angle not exceeding 30° to the horizontal m² **2** Upper surfaces inclined at 30°–60° to the horizontal m² **3** Surfaces inclined at an angle exceeding 60° to the horizontal m² **4** Curved surfaces m² **5** Domed surfaces m² **6** Surfaces of width not exceeding 300 mm m **7** Surfaces of width 300 mm–1 m m **8** Isolated groups of surfaces nr
4 Protective layers	**1** Sand asphalt **2** Flexible sheeting **3** Sand **4** Sand and cement screed **5** Tiles	**1** Upper surfaces inclined at an angle not exceeding 30° to the horizontal m² **2** Upper surfaces inclined at 30°–60° to the horizontal m² **3** Surfaces inclined at an angle exceeding 60° to the horizontal m² **4** Curved surfaces m² **5** Domed surfaces m² **6** Surfaces of width not exceeding 300 mm m **7** Surfaces of width 300 mm–1 m m **8** Isolated groups of surfaces nr
5 Sprayed or brushed waterproofing m²		
6 Sheet linings membrane	**1** Butyl rubber **2** EPDM rubber **3** PVC **4** Polypropylene **5** Polyethylene **6** Polyurethane **7** Proprietary system	**1** Upper surfaces inclined at an angle not exceeding 30° to the horizontal m² **2** Upper surfaces inclined at 30°–60° to the horizontal m² **3** Surfaces inclined at an angle exceeding 60° to the horizontal m² **4** Curved surfaces m² **5** Domed surfaces m² **6** Surfaces of width not exceeding 300 mm m **7** Surfaces of width 300 mm–1 m m **8** Isolated groups of surfaces nr

NOTE

Waterproofing may be measured by the number of *isolated groups of surfaces* of the same shape and dimensions instead of by the length or area of the separate surfaces.

MEASUREMENT RULES	DEFINITION RULES	COVERAGE RULES	ADDITIONAL DESCRIPTION RULES
M1 The areas measured shall be those of the surfaces covered. No deduction from the areas measured shall be made for holes and openings in the waterproofed surfaces each not exceeding 0.5 m^2 in area.		**C1** Items for waterproofing shall be deemed to include preparing surfaces, forming joints, overlaps, mitres, angles, fillets, built-up edges and laying to falls or cambers.	**A1** Item descriptions for waterproofing shall state the materials used and the number and thickness of coatings or layers.
M2 Surfaces of width not exceeding 1 m shall not be distinguished by inclination or by curvature. **M3** Waterproofing classed as to *curved* or *domed surfaces* shall not be distinguished by inclination.	**D1** Waterproofing shall be classed as to *curved* or *domed surfaces* only where a radius of curvature of the surface is less than 10 m.		
M4 *Isolated groups of surfaces* of different shape or dimensions shall be measured as separate items.	**D2** *Isolated groups of surfaces* shall be classed as such only where the total surface area of each group does not exceed 6 m^2.		**A2** Item descriptions for *isolated groups of surfaces* shall identify the work to be waterproofed and state its location.
M2 Surfaces of width not exceeding 1 m shall not be distinguished by inclination or by curvature. **M3** Waterproofing classed as to *curved* or *domed surfaces* shall not be distinguished by inclination.	**D1** Waterproofing shall be classed as to *curved* or *domed surfaces* only where a radius of curvature of the surface is less than 10 m.		
M4 *Isolated groups of surfaces* of different shape or dimensions shall be measured as separate items.	**D2** *Isolated groups of surfaces* shall be classed as such only where the total surface area of each group does not exceed 6 m^2.		**A2** Item descriptions for *isolated groups of surfaces* shall identify the work to be waterproofed and state its location.

CLASS X: MISCELLANEOUS WORK

Includes: **Fences, gates and their foundations**
Drainage to structures above ground
Rock filled gabions

FIRST DIVISION		SECOND DIVISION		THIRD DIVISION	
1 Fences	m	**1** Timber post and rail **2** Timber post and wire **3** Concrete post and wire **4** Metal post and wire **5** Coated metal post and wire **6** Timber close boarded **7** Metal guard rails **8** Road restraint system		**1** Height: not exceeding 1 m **2** 1–1.25 m **3** 1.25–1.5 m **4** 1.5–2 m **5** 2–2.5 m **6** 2.5–3 m **7** exceeding 3 m	
2 Gates and stiles	nr	**1** Timber gates **2** Metal gates **3** Stiles		**1** Width: not exceeding 1.5 m **2** 1.5–2 m **3** 2–2.5 m **4** 2.5–3 m **5** 3–4 m **6** 4–5 m **7** exceeding 5 m	
3 Drainage to structures above ground		**1** Mild steel **2** Cast iron **3** Plastics		**1** Gutters **2** Fittings to gutters **3** Downpipes **4** Fittings to downpipes	m nr m nr
4 Rock filled gabions		**1** Box of stated size **2** Mattress of stated thickness	nr m^2		
5 Open cell block systems	nr				

MEASUREMENT RULES	DEFINITION RULES	COVERAGE RULES	ADDITIONAL DESCRIPTION RULES
M1 Lengths of *fences* shall exclude lengths occupied by gates and stiles.	**D1** The heights used for classification of *fences* shall be measured from the Commencing Surface.	**C1** Items for *fences* shall be deemed to include excavation, preparation of surfaces, disposal of excavated material, upholding sides of excavation, backfilling, removal of existing services, concrete, formwork and reinforcement. **C2** Items for *fences* shall be deemed to include end posts, angle posts, straining posts and gate posts.	**A1** Item descriptions for *fences* which are erected to a curve of radius not exceeding 100 m or on a surface inclined at an angle exceeding 10° shall so state. **A2** The types and principal dimensions of *fences* and of their foundations shall be stated in item descriptions.
	D2 The width used for classification of *gates and stiles* shall be measured between the inside faces of posts.		**A3** The types and principal dimensions of *gates and stiles* shall be stated in item descriptions. **A4** Item descriptions for *gates* composed of more than one leaf shall identify the number of leaves in item descriptions.
	D3 *Fittings to gutters* shall include bends, angles, stop ends and outlets. *Fittings to downpipes* shall include bends, swan necks, shoes and roof outlets fixed directly to downpipes.	**C3** Items for *drainage to structures above ground* shall be deemed to include supports.	**A5** Item descriptions for *drainage to structures above ground* shall state the type, principal dimensions and materials of the components.
	D4 *Rock filled gabions* exceeding 300 mm thick shall be classed as box gabions and those not exceeding 300 mm thick as mattress gabions. Filling shall be deemed to be imported unless otherwise stated.		**A6** Item descriptions for *rock filled gabions* shall state the type and grading of filling, the size of mesh and the diameter of mesh wire. Details of protective coatings shall be stated.
			A7 The material, cross section type and principal dimensions or type shall be stated for *open cell block systems.*

CLASS Y: SEWER AND WATER MAIN RENOVATION AND ANCILLARY WORKS

Includes: **Preparation and renovation of existing sewers and water mains**
New manholes within the length of existing sewers
Work to existing manholes
Excludes: **Grouting carried out from outside the sewer (included in class C)**
New pipework (included in classes I–L)
New fittings and valves used in water main renovation (included in class J)
Extras to excavation and backfilling for new manholes and other chambers (included in class L)

FIRST DIVISION	SECOND DIVISION		THIRD DIVISION	
1 Preparation of existing sewers	**1** Cleaning	m		
	2 Removing intrusions	nr	**1** Laterals, bore not exceeding 150 mm **2** Laterals, stated profile and size exceeding 150 mm in one or more dimension **3** Other stated artificial intrusions	
	3 Closed-circuit television surveys	m		
	4 Plugging laterals, materials stated **5** Filling laterals and other pipes, materials stated	nr m^3	**1** Bore not exceeding 300 mm **2** Stated profile and size exceeding 300 mm in one or more dimension	
	6 Local internal repairs	nr	**1** Area: not exceeding 0.1 m^2 **2** 0.1–0.25 m^2 **3** stated exceeding 0.25 m^2	
2 Stabilization of existing sewers	**1** Pointing, materials stated **2** Pipe joint sealing, materials stated	m^2 nr		
	3 External grouting		**1** Number of holes **2** Injection of grout, materials stated	nr m^3

CLASS Y

MEASUREMENT RULES	DEFINITION RULES	COVERAGE RULES	ADDITIONAL DESCRIPTION RULES
M1 Lengths of sewers shall be measured along their centre lines between the inside surfaces of manholes but shall exclude lengths occupied by pipes and fittings comprising backdrops to manholes. **M2** Where work is expressly required to be carried out by excavation, crossings, reinstatement and other pipework ancillaries shall be measured in class K and extras to excavation and backfilling shall be measured in class L.		**C1** Items for work which is expressly required to be carried out by excavation shall be deemed to include preparation of surfaces, disposal of excavated material, upholding sides of excavation, backfilling and removal of dead services.	**A1** The location of the work in each item or group of items shall be stated so that the work can be identified by reference to the Drawings. **A2** Principal dimensions and profiles of sewers shall be stated in item descriptions. **A3** Work which is expressly required to be carried out manually or by remotely controlled methods shall each be so stated in item descriptions. **A4** Item descriptions shall identify work which is expressly required to be carried out by excavation and (except for manholes) shall state the maximum depth of excavation in stages of 1 m measured to the invert of the sewer or water main. **A5** Item descriptions for *preparation, stabilization, renovation* and *laterals* shall state the material forming the existing sewer or water main.
		C2 Items for *cleaning* shall be deemed to include making good resultant damage.	
	D1 Items shall be classed as *removing intrusions* where intrusions into the bores of existing sewers are to be removed prior to renovation.	**C3** Items for *removing intrusions* shall be deemed to include making good.	**A6** Item descriptions for *removing intrusions* shall state the materials forming the intrusions.
	D2 The areas stated in item descriptions for *local internal repairs* shall be the finished surface areas.	**C4** Items for *local internal repairs* shall be deemed to include cutting out and repointing.	
M3 No deduction shall be made from the areas of sewer surfaces measured for *pointing* for openings or voids each not exceeding 0.5 m^2 in area.		**C5** Items for *pointing* and *pipe joint sealing* shall be deemed to include preparation of joints.	
M4 *External grouting* shall be measured only where grouting is expressly required to be carried out as a separate operation from annulus grouting (Y 3 6 0).	**D3** *External grouting* shall be grouting of voids outside the existing sewer from within the existing sewer other than voids grouted in the course of annulus grouting.		**A7** Where *external grouting* is carried out through pipe joints, descriptions of items for the number of holes shall so state.

CLASS Y

FIRST DIVISION	SECOND DIVISION	THIRD DIVISION
3 Renovation of existing sewers	**1** Sliplining m	**1** Polyethylene **2** Polypropylene
	2 In situ jointed pipe lining m	**1** Polyethylene **2** Polypropylene **3** Glass reinforced plastic
	3 Segmental lining m	**3** Glass reinforced plastic **4** Glass reinforced concrete **5** Cast gunite **6** Resin concrete
	4 Stated proprietary lining m	
	5 Gunite coating of stated thickness m	
	6 Annulus grouting, materials stated m^3	
4 Laterals to renovated sewers	**1** Jointing nr	**1** Bore: not exceeding 150 mm **2** 150–300 mm **3** Stated profile and size exceeding 300 mm in one or more dimension
	2 Flap valves of stated size nr	**1** Remove existing **2** Replace existing **3** New flap valve of stated type
5 Water main renovation and ancillary works	**1** Cleaning m **2** Removing intrusions nr **3** Pipe sample inspections nr **4** Closed-circuit television surveys m **5** Cement mortar lining m **6** Epoxy lining m	**1** Nominal bore: not exceeding 200 mm **2** 200–300 mm **3** 300–600 mm **4** 600–900 mm **5** 900–1200 mm **6** stated exceeding 1200 mm
6 New manholes nr	**1** Brick **2** Brick with backdrop **3** In situ concrete **4** In situ concrete with backdrop **5** Precast concrete **6** Precast concrete with backdrop	**1** Depth: not exceeding 1.5 m **2** 1.5–2 m **3** 2–2.5 m **4** 2.5–3 m **5** 3–3.5 m **6** 3.5–4 m **7** stated exceeding 4 m
7 Existing manholes nr	**1** Abandonment	
	2 Alteration	
8 Interruptions h	**1** Preparation of existing sewers **2** Stabilization of existing sewers	
	3 Renovation of existing sewers	**1** Sliplining **2** In situ jointed pipe lining **3** Segmental lining **4** Stated proprietary lining **5** Gunite coating **6** Annulus grouting
	4 Work on laterals to renovated sewers **5** Work on manholes	

NOTE

Manholes may be measured in detail as set out in other classes of CESMM3.

MEASUREMENT RULES	DEFINITION RULES	COVERAGE RULES	ADDITIONAL DESCRIPTION RULES
			A8 Item descriptions for *slip-lining, in situ jointed pipe lining, segmental lining* and *stated proprietary lining* shall state the type of lining, its minimum finished internal size and its thickness or grade. **A9** Item descriptions for *in situ jointed pipe lining* and *segmental lining* shall state the offset where the lining is curved to an offset which exceeds 35 mm per metre.
M5 The volume measured for *annulus grouting* shall not include the volume measured for external grouting (Y 2 3 2).	**D4** *Annulus grouting* shall be grouting of the annular voids between new linings and existing sewers and of other voids grouted in the course of grouting annular voids.		
		C6 Items for *laterals* shall be deemed to include the work involved in connecting to the lining within 1 m from the inside face of the lined sewer.	**A10** Item descriptions for *jointing* laterals shall state the type of lining to which the laterals are to be connected and identify those laterals which are to be regraded.
M6 Lengths of water mains shall be measured along their centre lines and shall include lengths occupied by fittings and valves.	**D5** *Pipe sample inspections* and *closed-circuit television surveys* shall include work carried out either before or after cleaning and lining.	**C7** Items for *pipe sample inspections* shall be deemed to include replacing the length removed by new pipework.	**A11** Item descriptions for *lining* shall state the materials, nominal bores and thicknesses of the lining.
	D6 The depth of *manholes* shall be measured from the tops of covers to channel inverts or tops of base slabs, whichever is the lower.	**C8** Items for *new manholes* shall be deemed to include excavation, preparation of surfaces, disposal of excavated material, upholding sides of excavation, backfilling, concrete, reinforcement, formwork, joints, finishes and reinstatement. **C9** Items for *new manholes* shall be deemed to include all items of metalwork and pipework which occur within or at the surface of the manhole. **C10** Items for *new manholes* with backdrops shall be deemed to include the pipes and fittings comprising the backdrop. **C11** Items for *new manholes* which replace existing manholes shall be deemed to include breaking out and disposal of existing manholes.	**A12** Type or mark numbers shall be stated in item descriptions for *manholes* of which details are given elsewhere in the Contract. Item descriptions shall identify different configurations of manholes. **A13** Types and loading duties of covers shall be stated in item descriptions for *new manholes.* **A14** Item descriptions shall separately identify *new manholes* which replace existing manholes. **A15** Item descriptions for *existing manholes* shall state details of the work required.
M7 *Interruptions* shall be measured only where a minimum pumping capacity is expressly required and for periods of time during normal working hours when the flow in the sewer exceeds the installed pumping capacity and work is interrupted.			

CLASS Z: SIMPLE BUILDING WORKS INCIDENTAL TO CIVIL ENGINEERING WORKS

Includes: **Carpentry and joinery**
Insulation
Windows, doors and glazing
Surface finishes, linings and partitions
Piped building services
Ducted building services
Cabled building services

FIRST DIVISION		SECOND DIVISION		THIRD DIVISION	
1 Carpentry and joinery		**1** Structural and carcassing timber		**1** Floors **2** Walls and partitions **3** Flat roofs **4** Pitched roofs **5** Plates and bearers **6** Struts **7** Cleats **8** Trussed rafters and roof trusses	m m m m m m nr nr
		2 Strip boarding **3** Sheet boarding	m^2 m^2	**1** Floors **2** Sloping upper surfaces **3** Walls **4** Soffits	
		4 Stairs and walkways	nr	**1** Stairways and landings **2** Walkways and platforms **3** Isolated balustrades **4** Ladders	
		5 Miscellaneous joinery	m	**1** Skirtings **2** Architraves **3** Trims **4** Shelves	
		6 Units and fittings	nr	**1** Base units **2** Wall units **3** Work tops **4** Notice-boards	
2 Insulation	m^2	**1** Sheets **2** Quilts **3** Boards **4** Loose fill		**1** Floors **2** Sloping upper surfaces **3** Walls **4** Soffits	
3 Windows, doors and glazing		**1** Timber **2** Metal **3** Plastics	nr nr nr	**1** Windows **2** Window sub-frames **3** Doors **4** Frames or lining sets **5** Screens and borrowed lights **6** Roof lights	
		4 Ironmongery	nr	**1** Hinges **2** Door closers **3** Locks **4** Bolts **5** Handles **6** Plates **7** Brackets	

CLASS Z

Excludes: **Drainage below ground (included in classes I–L)**
Ducts and trenches for electrical services below ground or outside the building (included in classes I–L)
Metalwork (included in class N)
Civil engineering timber works (included in class O)
Brickwork, blockwork and masonry (included in class U)
Painting (included in class V)
Asphalt work (included in class W)
Roofing, cladding and coverings (included in class W)
Drainage to structures above ground (included in class X)

<table>
<tr><th>MEASUREMENT RULES</th><th>DEFINITION RULES</th><th>COVERAGE RULES</th><th>ADDITIONAL DESCRIPTION RULES</th></tr>
<tr>
<td>**M1** The lengths and areas measured for *carpentry and joinery* items shall be measured net with no allowance for joints or laps.

M2 No deduction from the areas measured shall be made for holes and openings each not exceeding 0.5 m^2 in area.

M3 *Struts* between joists shall be measured on plan across joists.</td>
<td rowspan="2">**D1** Sizes stated in item descriptions for *carpentry and joinery* shall be nominal sizes unless otherwise stated.

D2 Ceiling joists shall be classed as *pitched roofs* where they are associated with pitched roofs.

D3 Timber and other manufactured boards shall be classed as *sheet boarding*.

D4 *Boarding and insulation* shall be classified according to their angle of inclination as follows:
<table><tr><th>Class</th><th>Angle of inclination to the vertical</th></tr><tr><td>Floors</td><td>75°–90°</td></tr><tr><td>Sloping upper surfaces</td><td>15°–75°
Not exceeding 15°</td></tr><tr><td>Walls</td><td>15°–90°</td></tr><tr><td>Soffits</td><td></td></tr></table>
Surfaces of columns shall be classed as *walls*. Surfaces of beams shall be classed as *soffits*.</td>
<td>**C1** Items for *carpentry and joinery* shall be deemed to include fixing, boring, cutting, jointing, supply of fixing components and drilling or cutting associated work.</td>
<td>**A1** Item descriptions for *carpentry and joinery* shall state the materials used and identify whether they are sawn or wrought and any treatment, selection or protection for subsequent treatment.

A2 Item descriptions shall state the nominal thickness of *boarding*.

A3 The width of *boarding* shall be stated in item descriptions in 100 mm stages where it does not exceed 300 mm.

A4 Item descriptions for *structural and carcassing timber* (other than *cleats, rafters and trusses*) and *miscellaneous joinery* shall state the overall nominal gross cross-sectional dimensions and the number of different cross-section shapes where there are more than one.

A5 Item descriptions shall identify applied laminates and coverings.

A6 Item descriptions shall identify the shape, size and limits of *stairs and walkways* and *units and fittings*.</td>
</tr>
<tr><td></td><td></td><td>**A7** Item descriptions for *insulation* shall state the materials and their overall nominal thickness.</td></tr>
<tr><td></td><td></td><td rowspan="2">**C2** Items for *windows, doors and glazing* shall be deemed to include fixing, supply of fixing components and drilling or cutting of associated work.</td><td>**A8** Item descriptions for *windows, doors and glazing* (Z 3 1-3 *) shall identify the shape, size and limits of the work.</td></tr>
<tr><td></td><td></td><td>**A9** Materials shall be stated in item descriptions for *ironmongery*.</td></tr>
</table>

CLASS Z

FIRST DIVISION	SECOND DIVISION		THIRD DIVISION	
	5 Glazing		**1** Glass	m²
			2 Glass in large panes	m²
			3 Special glass	nr
			4 Hermetically sealed units	nr
			5 Mirrors	nr
	6 Patent glazing	m²	**1** Roofs	
			2 Opening lights	
			3 Vertical surfaces	
4 Surface finishes, linings and partitions	**1** In situ finishes, beds and backings		**1** Floors	m²
	2 Tiles		**2** Sloping upper surfaces	m²
	3 Flexible sheet coverings		**3** Walls	m²
	4 Dry partitions and linings		**4** Soffits	m²
			5 Surfaces of width not exceeding 300 mm	m
			6 Surfaces of width 300 mm–1 m	m
	5 Suspended ceilings		**1** Depth of suspension: not exceeding 150 mm	m²
			2 150–500 mm	m²
			3 exceeding 500 mm	m²
			4 Bulkheads	m
			5 Access panels	nr
			6 Fittings	nr
	6 Raised access floors	m²		
	7 Proprietary system partitions		**1** Solid	m
			2 Fully glazed	m
			3 Partially glazed	m
			4 Door units	nr
	8 Framed panel cubicle sets	nr		

<table>
<tr><th>MEASUREMENT RULES</th><th>DEFINITION RULES</th><th>COVERAGE RULES</th><th>ADDITIONAL DESCRIPTION RULES</th></tr>
<tr>
<td></td>
<td>D5 Panes which exceed 4 m² shall be classed as large panes.

D6 The following materials used in glazing shall be classed as special glass
(a) glass whose thickness exceeds 10 mm
(b) non-rectangular panes of glass
(c) toughened, laminated, solar control and other speciality glass
(d) acrylic, polycarbonate and similar materials.</td>
<td rowspan="2">C2 Items for windows, doors and glazing shall be deemed to include fixing, supply of fixing components and drilling or cutting of associated work.</td>
<td>A10 Item descriptions for glazing shall identify the materials, their nominal thicknesses, the method of glazing and the method of securing the glass.

A11 Item descriptions shall identify the construction of hermetically sealed units.

A12 Item descriptions for special glass, hermetically sealed units and mirrors shall identify the shape and size of panes.

A13 Item descriptions for glazing shall identify work which is curved.</td>
</tr>
<tr>
<td>M2 No deduction from the areas measured shall be made for holes and openings each not exceeding 0.5 m² in area.

M4 The areas measured for patent glazing shall include areas occupied by glazing bars.</td>
<td></td>
<td>A14 Item descriptions shall identify the shape, size and limits of patent glazing.

A15 Incidental metalwork supporting patent glazing shall be identified in item descriptions.

A16 Item descriptions for patent glazing shall identify work which is curved.</td>
</tr>
<tr>
<td>M2 No deduction from the areas measured shall be made for holes and openings each not exceeding 0.5 m² in area.

M5 The length measured for proprietary system partitions shall exclude the length of voids which extend to the full height of the partition.</td>
<td>D7 Surface finishes, linings and partitions (Z 4 1-4 *) exceeding 1 m wide shall be classified according to their angle of inclination as follows:
<table>
<tr><th>Class</th><th>Angle of inclination to the vertical</th></tr>
<tr><td>Floors</td><td>75°–90°</td></tr>
<tr><td>Sloping upper surfaces</td><td>15°–75°</td></tr>
<tr><td></td><td>Not exceeding 15°</td></tr>
<tr><td>Walls</td><td>15°–90°</td></tr>
<tr><td>Soffits</td><td></td></tr>
</table>
Surface of columns exceeding 1 m wide shall be classed as walls. Surfaces of beams exceeding 1 m wide shall be classed as soffits.</td>
<td>C3 Items for surface finishes, linings and partitions shall be deemed to include fixing, supply of fixing components and drilling or cutting of associated work.

C4 Items for surface finishes, linings and partitions shall be deemed to include preparing surfaces, forming joints, mitres, angles, fillets, built-up edges and laying to cambers or falls.

C5 Items for surface finishes, linings and partitions shall be deemed to include forming holes, cutting and making good for services.

C6 Items for suspended ceilings shall be deemed to include associated primary support systems and edge trims.</td>
<td>A17 The materials, surface finish and finished thickness shall be identified in item descriptions for surface finishes, linings and partitions.

A18 Lathing and baseboarding associated with in situ finishes, bed and backings shall be identified in item descriptions.

A19 The girth of bulkheads shall be identified in item descriptions.

A20 The overall dimensions of access panels and fittings in ceilings, door units in proprietary system partitions and framed panel cubicle sets shall be identified in item descriptions.

A21 Item descriptions for suspended ceilings shall state the depth of the suspension where it exceeds 500 mm in stages of 500 mm.</td>
</tr>
</table>

CLASS Z

FIRST DIVISION	SECOND DIVISION		THIRD DIVISION	
5 Piped building services	**1** Pipework		**1** Pipes **2** Fittings **3** Insulation	m nr m
	2 Equipment	nr	**1** Boiler plant and ancillaries **2** Convectors and radiators **3** Pumps **4** Cisterns and tanks	
	3 Sanitary appliances and fittings	nr		
6 Ducted building services	**1** Circular ductwork **2** Rectangular ductwork		**1** Straight **2** Curved **3** Fittings **4** Insulation	m m nr m
	3 Equipment	nr	**1** Conditioning and handling units **2** Heaters **3** Fans **4** Filters	

MEASUREMENT RULES	DEFINITION RULES	COVERAGE RULES	ADDITIONAL DESCRIPTION RULES
M6 Lengths of pipes shall be measured along their centre lines and shall include lengths occupied by *fittings*.	**D8** Taps shall be classed as pipe *fittings*.	**C7** Items for *piped services systems* shall be deemed to include fixing and supply of fixing components. **C8** Items for *piped services systems* shall be deemed to include commissioning.	**A22** The location or type of *piped building services* in each item or group of items shall be stated in item descriptions so that the work included can be identified by reference to the Drawings. **A23** The materials, joint types, and nominal bores of *pipework* shall be stated in item descriptions and reference given to applicable specifications and specified qualities. *Fittings* on pipes of different nominal bores shall be identified by their larger bore. **A24** The type of *equipment* or *sanitary appliances and fittings* and the materials, size or capacity, and the method of fixing shall be stated in item descriptions and reference given to applicable specifications and specified qualities. **A25** Traps shall be identified in item descriptions for *sanitary appliances and fittings*.
M7 Lengths of *ducted services systems* shall be measured along their centre lines and shall include lengths occupied by *fittings*.		**C9** Items for *ducted building services* shall be deemed to include fixing and supply of fixing components. **C10** Items for *ducted building services* shall be deemed to include commissioning.	**A26** The location or type of *ducted building services* in each item or group of items shall be stated in item descriptions so that the work included can be identified by reference to the Drawings. **A27** The materials, joint types, and size of *ductwork* shall be stated in item descriptions and reference given to applicable specifications and specified qualities. *Fittings* on ducts of different sizes shall be identified by their largest size. **A28** The type of *equipment* and the materials, size or capacity, and the method of fixing shall be stated in item descriptions and reference given to applicable specifications and specified qualities.

CLASS Z

FIRST DIVISION	SECOND DIVISION		THIRD DIVISION	
7 Cabled building services	**1** Cables	m	**1** Laid or drawn into conduits, trunking or ducts **2** Laid on trays **3** Fixed to surfaces **4** Laid in trenches **5** Suspended	
	2 Conduits		**1** Plain **2** Flexible **3** Box fittings	m m nr
	3 Trunking **4** Busbar trunking **5** Trays		**1** Plain **2** Fittings	m nr
	6 Earthing and bonding		**1** Tapes **2** Fittings	m nr
	7 Final circuits	nr	**1** Cable only **2** Cable and conduit	
	8 Equipment and fittings	nr	**1** Equipment **2** Switches **3** Lighting outlets **4** Sockets	

NOTE

Windows and their *sub-frames (Z 3 1-3 1-2), doors* and their *frames or lining sets* (Z 3 1-3 3-4) and the associated *ironmongery* may be included in the measurement of *doors* or *windows* provided that the work included is identified in item descriptions in accordance with paragraph 5.12 and the appropriate statements are given in the Preamble in accordance with paragraph 5.4.

NOTE

Taps may be included in item descriptions for *sanitary appliances and fittings* instead of by separate items for pipe *fittings* provided that the work included is identified in item descriptions in accordance with paragraph 5.12 and the appropriate statements are given in the Preamble in accordance with paragraph 5.4.

MEASUREMENT RULES	DEFINITION RULES	COVERAGE RULES	ADDITIONAL DESCRIPTION RULES
M8 Lengths measured for *cables* shall exclude lengths occupied in sags and tails. **M9** Lengths measured for *conduits, trunking, busbar trunking* and *cable trays* shall include lengths occupied by fittings. **M10** Lengths measured for *cables laid or drawn into conduits, trunking or ducts* or *laid on trays* shall be measured as the length of conduit, trunking, duct or tray.	**D9** Simple low voltage power and lighting circuits which loop from distribution boards shall be classed as *final circuits*.	**C11** Items for *cabled building services* shall be deemed to include determining circuits, terminations and connections, providing draw wires and draw cables, cleaning trunking, ducts and trays and threading cables through sleeves. **C12** Items for *cabled building services* shall be deemed to include fixing and supply of fixing components. **C13** Items for *cabled building services* shall be deemed to include commissioning. **C14** Items for *conduits* shall be deemed to include fittings other than *boxed fittings*.	**A29** The location or type of *cabled building services* in each item or group of items shall be stated in item descriptions so that the work included can be identified by reference to the Drawings. **A30** The materials, size or capacity, and method of fixing shall be identified in item descriptions for *cabled building services* (Z 7 1-7 *) and reference given to applicable specifications and specified qualities. **A31** The type and size or capacity of *equipment and fittings* (Z 7 8 *) shall be stated in item descriptions and reference given to applicable specifications and specified qualities.